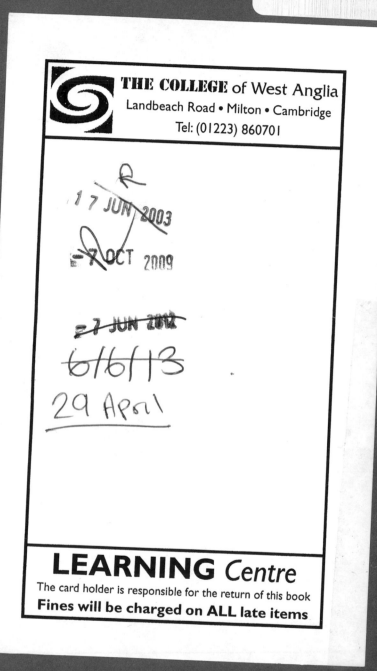

Birds of Prey
their biology and ecology

Birds of Prey

their biology and ecology

Leslie Brown

HAMLYN
London · New York · Sydney · Toronto

Acknowledgements

Colour plates
AQUILA PHOTOGRAPHICS: W S Paton 198–199 top, Eric Soothill/Peter Whitehead 35 top, M Wilkes 38 top; ARDEA PHOTOGRAPHICS: Peter Alden 87 top right, M D England 198 bottom, K Fink 94 top, Clem Haagner 47 top right, Don Hadden 42–43, Peter Steyn 81 bottom; C H BROWN: 38 bottom, 131 top; L H BROWN: 82–83 bottom, 86–87, 134 bottom, 134–135 top, 199 bottom; BRUCE COLEMAN LTD: Dennis Green 138 bottom, Hans Reinhard 207, H Rivarola 83 top; ERIC HOSKING: 90 top, 194 bottom; JACANA: J L S Dubois 135 bottom, Dupont 195 top, Hervy 82 top, Robert 39 bottom, 130, 131 bottom, Juan Solaro 95 top, J P Varin 34 bottom; FRANK W LANE: R Austing 94 bottom, 95, 206, A Christiansen 34 top, Norman Duerden 139, P J Ginn 143 bottom, Heinz Schrempp 39 top, Ronald Thompson 195 bottom, D Zingel 90 bottom, W T Miller 194 top; NATURAL HISTORY PHOTOGRAPHIC AGENCY: Anthony Bannister 46–47, F V Blackburn 143 top, J Good 138 top, James Tallon 142; NATURFOTO: Sigurd Halvorsen 202 bottom, Karl Holgard 91 top, Karl Holgard/B Gensbol 203, Carlo Jensen 35 bottom; WORLD WILDLIFE FUND: W A Newlands 202 top right.

Front jacket: BRUCE COLEMAN LTD: H M Barnfather
Back jacket: ARDEA PHOTOGRAPHICS: T A Willock

Black and white photographs
AQUILA PHOTOGRAPHICS: D Green 73, 225, M Wilkes 115; ARDEA PHOTOGRAPHICS: P Blasdale 85, R J C Blewitt 170 bottom, 215, M D England 171 top, K Fink 185, Clem Haagner 53, 54, 109, 112, 210, 214, Ake Lindau 11, Richard Vaughan 133, 169, Adrian Warren 220, J S Wightman 208 top, T A Willock 175; L H BROWN: 48, 173, 176, 178, 179, 180, 208 bottom; FORESTRY COMMISSION: 228; ERIC HOSKING: 92, 170 top, 171 bottom, 177; JACANA: Serge Chevallier 116, Pierre Petit 209; H KRUUK: 212; FRANK W LANE: Ronald Austing 14, 168, A Christiansen 79, John Karmali 16, W T Miller 211, Georg Quedens 71; B U MEYBURG: 174, 229; NATURAL HISTORY PHOTOGRAPHIC AGENCY: Anthony Bannister 184, E Hawumautha Rao 13; NATURFOTO: Sigurd Halvorsen 111, Arne Schmitz 146; ROYAL SOCIETY FOR THE PROTECTION OF BIRDS: Michael W Richards 223; ROB AND BESSIE WELDER WILDLIFE FOUNDATION: Dr Eric G Bolen 224.

Title spread: ARDEA PHOTOGRAPHICS: Hans and Judy Beste

Line drawings by Ian Willis

For Dean and Tavvy

Published by The Hamlyn Publishing Group Limited
London · New York · Sydney · Toronto
Astronaut House, Feltham, Middlesex, England
Copyright © The Hamlyn Publishing Group Limited 1976

ISBN 0 600 31306 9

Phototypeset by Photoprint Plates Limited, Rayleigh, England
Colour separations by Culver Graphics Litho Limited, Lane End, Buckinghamshire, England
Printed in Spain by Mateu Cromo, Madrid

Contents

Classification and distribution

Any bird that preys on other living creatures is, strictly speaking, a predator, that is, a bird of prey. So broad a definition, however, would include everything from tiny, insectivorous warblers to ponderous pelicans, most land birds, and practically all seabirds. Few bird species are entirely vegetarian, and many seed eaters, such as finches, feed their young on insects. Birds of prey require more precise definition.

The true birds of prey are those which have powerful, taloned feet for grasping and killing, and hooked beaks for tearing flesh. They are included in two not very closely related orders of birds: the Falconiformes, including kites, vultures, hawks, eagles, and falcons; and the Strigiformes, including all owls. Birds of these orders superficially resemble one another in their hooked beaks and taloned feet, but the Falconiformes are most nearly related to ducks and gamebirds, while the owls are placed between nightjars and cuckoos. There is some anatomical and behavioural evidence to suggest that the long-winged falcons may be more closely related to owls than was thought, but the resemblances between these groups are probably due to parallel evolution, whereby species not very closely related develop similar structures through the performance of similar functions. There are many examples of this process on a lesser scale among the Falconiformes themselves.

The Falconiformes are generally diurnal, that is, they hunt by day, though some overlap with dusk-hunting owls, while some owls are largely or partly diurnal. Falconiformes hunt mainly when they can see, so that good eyesight is crucial to their lives. The owls, hunting by night, depend more on their ears, as you or I would in the dark. Both have developed similar killing structures – the taloned feet, and similar hooked beaks for eating – but find and hunt their food in different ways. Some diurnal birds of prey – the New and Old World vultures – do not now kill, so that their talons have atrophied, though they retain the hooked beak for eating flesh.

We can now forget about owls, except for an occasional comparative mention. This book concerns the Falconiformes, or diurnal birds of prey; and there is more than enough to say about them. Many books have already been written about them, so that here we explore in some detail their relationships with their habitats, and the adaptations and modes of predation they have developed in response both to habitats and to the type of prey they eat – in so far as these subjects have been studied at all. More questions may be posed than can be answered, showing that there is still abundant scope for research by anyone interested.

The order Falconiformes occurs worldwide, on every continent except Antarctica, and on many small oceanic islands. The 287 accepted species vary from tiny, insectivorous falconets weighing 60 grams (little more than a Song Thrush), to huge eagles, griffon vultures, and condors weighing 7 to 9 kilograms. These include some of the largest flying birds, and the Harpy Eagle is certainly the most formidable living bird. Falconiforms feed on everything from beetles and termites to dead elephants and whales. They live in almost every known habitat, from Arctic tundra to deserts, and from open steppe to deep tropical forest. They hunt mainly by sight, occasionally aided by ear, but very rarely (in a few New World vultures) by smell. Owls are almost equally widespread, so that in any varied habitat some species is usually present to eat many of the other animals occurring there, by day or night, dead or alive. Owls, however, rarely or never eat large dead animals because these do not move and make a noise locatable by ear at night. Carrion is the prerogative of vultures and some others which find it by sight by day; or of mammalian scavengers such as hyenas which locate it by smell at night.

Any regional handbook on birds shows that there is no general agreement as to how the 287 diurnal birds of prey should be classified. No two authors of such standard works agree in general or in detail. This is making confusion about classification worse confounded than need be, however, because two rather different modern classifications are available, and for many years a generally acceptable basis has been provided by the Peters' list of *Birds of the World*. Those who ignore such basic lists and insist, without explanation, on their own order of classification do a disservice to ordinary birdwatchers by obscuring much basic work already done.

The two modern classifications are those of the late Dr Erwin Stresemann and of Dr Dean Amadon. In the Stresemann classification, which departs radically from the Peters' list, the list of families and genera is based largely, but not entirely, on the order of moulting of the nine primary feathers of the wing, defined in three ways: 1 **ascendant**, beginning with primary number 4 and proceeding inwards and outwards from there; 2 **descendant**, beginning with the innermost primary, number 1, and working outwards to the wingtip; 3 **irregular**, with no clear sequence. The third mode can be further subdivided into species that moult **descendant** as juveniles but

Fig. 1 Three modes of wing moult in Falconiformes: (a) ascendant (Gyrfalcon) – moult begins at primary 4 and proceeds both outwards to wingtip and inwards towards secondaries; (b) descendant (Goshawk) – moult begins at primary 1 and proceeds outwards to wingtip; (c) irregular (Turkey Vulture) – moult begins at several centres and proceeds both outwards and inwards from these.

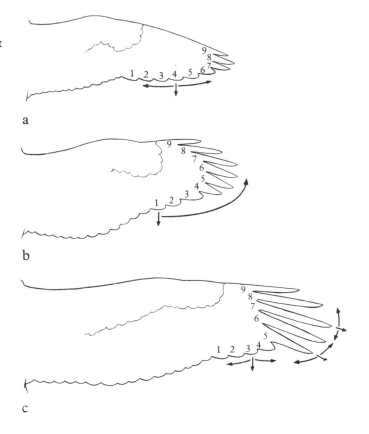

irregular as adults, and those that moult **irregularly** at all ages. Descendant moult is considered more primitive, and irregular moult derived from it. See Fig. 1.

Classification based on this system shares some common ground with the Amadon classification which is based on the Peters' list revised in the light of recent research. In both, the New World vultures are considered to be a sub-order, while the unique Secretary Bird is placed in its own suborder or at least superfamily. In Stresemann's arrangement, however, the kites, honey buzzards, and sea eagles (which are regarded by Peters and Amadon as relatively primitive) come late in the evolutionary sequence and are apparently considered more advanced than the booted eagles with feathered tarsi, while the rather specialized Old World vultures are placed close to the primitive New World vultures. Also most species that moult irregularly are large, and irregular moult may be more closely connected with weight and wing-loading than with genetic origin.

Anyone specially interested in classification should study Stresemann's interesting and original arrangement. The Amadon classification, however (first proposed in *Eagles, Hawks and Falcons of the World*, 1969), is more likely to become standard and has already been accepted and used in some modern works. In this classification the living and

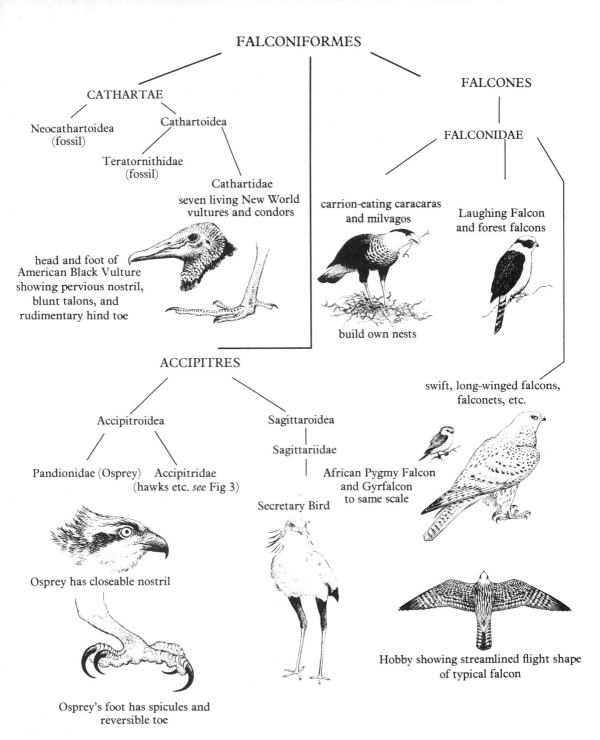

FALCONIFORMES

CATHARTAE

Neocathartoidea
(fossil)

Cathartoidea

Teratornithidae
(fossil)

Cathartidae
seven living New World
vultures and condors

head and foot of
American Black Vulture
showing pervious nostril,
blunt talons, and
rudimentary hind toe

FALCONES

FALCONIDAE

carrion-eating caracaras
and milvagos

Laughing Falcon
and forest falcons

build own nests

ACCIPITRES

Accipitroidea

Sagittaroidea

Sagittariidae

Pandionidae (Osprey)   Accipitridae
                       (hawks etc. *see* Fig 3)

African Pygmy Falcon
and Gyrfalcon
to same scale

swift, long-winged falcons,
falconets, etc.

Secretary Bird

Osprey has closeable nostril

Osprey's foot has spicules and
reversible toe

Hobby showing streamlined flight shape
of typical falcon

Fig. 2 The broad basis of the
Amadon classification.

9

known fossil birds of prey are divided into three suborders:

1 the Cathartae, with two superfamilies, Neocathartoidea and Cathartoidea, including the living New World vultures and some fossil condors;

2 the Accipitres, with two superfamilies, the Accipitroidea including the Osprey and all living hawks, eagles, and Old World vultures; and the Sagittaroidea including only the Secretary Bird;

3 the Falcones, with a single family, the Falconidae, including all living and fossil falcons, caracaras, and milvagos. Falconidae differ from all other birds of prey in many ways, but all moult ascendant. Although they are the strongest argument for a classification based on moult, Stresemann considers falcons and their allies only a family, whereas Amadon elevates them to a suborder. See Fig. 2.

1 The suborder Cathartae includes seven living species of New World vultures and condors, and some fossils. All living species and most fossils occur only in the Americas but some very primitive fossils, perhaps cathartids, have been found in Europe. Cathartids include the largest living species, the Andean Condor.

The two superfamilies of Carthartae are: (a) the Neocathartoidea, only known from a fossil species, *Neocathartes grallator*, which could not fly or was entirely terrestrial; (b) the Cathartoidea, including two families, the Cathartidae comprising the seven living New World vultures and condors and some fossils, and the Teratornithidae or teratorns, represented by huge fossils found mainly in the tar pits of La Brea. The Pleistocene *Teratornis merriami*, which existed only about 10 000 years ago, had a wingspan of more than 4 metres and weighed 20 kilograms, twice the weight of an Andean Condor, but it was tiny beside the much larger *Teratornis gigans*, known only from a foot-bone. We may assume these huge birds died out because they were just too big to cope with post-Pleistocene food shortages but it is a wonder they were able to fly at all.

The living New World vultures differ from other known birds of prey in their pervious nostrils – a thin rod can be thrust right through the beak; a rudimentary hind toe – their feet resemble those of chickens rather than eagles'; large olfactory chambers, suggesting that they have a sense of smell, unique in diurnal birds of prey; none of them builds a nest and they all lay eggs which, when held up to the light, have a yellowish inside shell. They are obviously very different from all others, and no close relationships with the next suborder, Accipitres, are demonstrable. They form an ancient group which has now become isolated in the New World, but perhaps once lived in Europe.

2 The suborder Accipitres is by far the largest and most varied of the three, including 220 species, distributed

worldwide, and divided into two superfamilies, the Accipitroidea and the Sagittaroidea, sharing certain common characters. All lay eggs with a greenish inside shell, build their own nests, and all those that have been studied squirt their droppings some distance (except some snake eagles and the Lammergeier which produce hard, dry droppings which they cannot squirt). They are, however, so immensely varied that they deserve detailed examination.

The superfamily Accipitroidea is divided into two families: the Pandionidae, with only one species, the Osprey, *Pandion haliaetus*; and the Accipitridae, with 218 species. There is controversy about precisely how the Osprey is related to other birds of prey and it has often been placed between the end of the list of hawks and eagles and the caracaras. It is unique in having a reversible outer toe (a device probably associated with grasping the fish it eats) and a nostril it can close at will, but it builds its own nest and lays eggs with a greenish inside shell. Dean Amadon considered that it ought to be placed early in the list because of certain anatomical features shared with rather primitive kites, notably lack of a bony eye-shield, and a similar sternum. It is arguable whether the Osprey should be regarded as near to primitive kites, or highly specialized by evolution and so better placed late in the list. Clearly, however, it should be in its own family or subfamily, wherever that is placed. It is a very well-known, much-studied, nearly cosmopolitan species but further work is needed to establish its precise relationships.

The 218 species in the family Accipitridae are extremely varied and in many old classifications were divided into a number of recognizable subfamilies. It is a pity that this convenient arrangement has been abandoned, because it certainly helped to clarify relationships for most people, even if there are borderline cases or intermediate forms which are now thought to make such subdivisions

An Osprey in flight; the systematic position of these unique fish-eating raptors is uncertain, but some features such as the obvious bend in the wing suggest alliance to kites.

untenable. The subfamilies are noted in the species list (Appendix I). Accipitridae as a whole occur worldwide, but some old subfamilies were confined to the New or Old World, or most characteristic of one continent.

The broad groupings forming the old subfamilies are: Perninae, including honey buzzards and aberrant kites – most lack a bony eyeshield and include some extremely specialized forms such as the wasp-eating Honey Buzzard (not really a buzzard); Machaeramphinae including only the unique, crepuscular, bat-eating Bat Hawk *Machaerhamphus*; Elaninae, including the well-known Black-shouldered Kite; Milvinae, including the highly specialized snail kites, *Rostrhamus*, the true carrion or scavenging kites and the eastern Brahminy Kites. Although these three groups of kites are very varied, and some are highly specialized, they seem related to one another and the Brahminy Kite appears to link them to the Haliaeetinae, including sea and fish eagles. These have developed a bony eyeshield (which they may need as active and powerful predators) but they resemble true kites in several other ways, notably in display, piratical habits, and nuptial and moult patterns. They are more doubtfully allied through the extraordinary vegetarian Vulturine Fish Eagle, *Gypohierax angolensis*, to the Old World vultures, Aegypiinae. The connecting link may be the Egyptian Vulture, *Neophron percnopterus*, which is black and white like some fish eagles, but differs in many way from *Gypohierax*. This is the first apparently weak link in the linear chain of relationships.

In a linear classification sequence or list, the Old World vultures are followed by the snake and serpent eagles, Circaetinae. These are obviously not closely related to Old World vultures, but are a highly specialized reptile- and frog-eating group. They are placed here for want of any better arrangement. They are probably related to the Harrier and Crane Hawks, Polyboroidinae, which have unique, double-jointed legs, but the similarities may be more superficial than real. The Polyboroidinae are, in turn, apparently related to the harriers, Circinae, a remarkably uniform group of long-winged, long-tailed, long-legged hawks, all hunting by buoyant, flapping flight in open country, and breeding on the ground (except one which breeds in low trees). The harriers are apparently rather weakly related to the chanting goshawks, *Melierax*, and then to the true sparrowhawks and goshawks, Accipitrinae. These are all short-winged, long-tailed species of woodland or forest, and include the largest genus of all, the true hawks, *Accipiter* spp, with forty-seven species. These are fairly clearly linked through intermediate forms sometimes called sub-buteonines, to buzzards, *Buteo* spp, and buteonine eagles, Buteoninae. These vary from quite small species to the gigantic Harpy Eagle which, with its

An Indian Black Kite; probably the commonest and most successful raptor in the world, partly because it has learned to live with man.

near allies in New Guinea and the Philippines, apparently are a relict group and the culmination of one line of evolution. The final subfamily includes the booted eagles with feathered tarsi, Aquilinae, which share many habits with buzzards but differ in some respects. They culminate in the magnificent Crowned and Martial Eagles of Africa.

Some subfamilies of these very varied hawks, eagles, and vultures are confined to certain parts of the world. The Old World vultures, and snake eagles are confined to the Old World, excluding Australasia. The true scavenging kites and Brahminy Kites occur only in the Old World but reach Australasia. The fish and sea eagles are mainly Old World but include the North American Bald Eagle; oddly, none occurs in South America. The Harrier and Crane Hawks occur in tropical Africa and South America and suggest relict links between these continents. The Buteoninae occur worldwide except for Australasia.

The remainder occur worldwide but may be more numerous and varied in some continents. Among the booted eagles, Aquilinae, twenty-five are Old World, and only five New World species. Among the Buteoninae, the greatest variety – thirty-seven species in ten genera – are found in Central and South America, and buteos themselves are more varied in temperate North American forests than European forests. In Australasia, where there are no buteos or Old World vultures, the kites have developed large, powerful, buzzard-like forms, and the Wedge-tailed Eagle partially fills the vulture niche.

The ramifications of the accipitrine family tree are best

A Goshawk. An example of the largest species in the largest genus of birds of prey, *Accipiter*, widespread in all northern woodlands.

shown in the form of a diagram (Fig. 3). According to the Amadon classification, two main lines of evolution may have started from primitive, insectivorous kites, one culminating in sea eagles and Old World vultures (where the linear chain or list has a clear, definite break), and the other leading to snake eagles, harriers, accipiters, buteos, and booted eagles. This rather rational arrangement is the best we can do, but because the hawks and eagles comprise a very old group of birds, gaps are inevitable.

The unique Secretary Bird is placed in a superfamily of its own, the Sagittaroidea, with one family, Sagittariidae, and one genus and species, *Sagittarius serpentarius*. Some even doubt that this is a true falconiform; it may be related more closely to the South American cariamas, and through them to cranes and bustards. Little seems to be known about cariamas, so that for the present the Secretary Bird is considered a unique and highly aberrant, semiterrestrial, long-legged falconiform, not very closely related to anything else, but building its own nest, having an undulating display flight like many Accipitres, laying eggs with a greenish inside shell, and squirting its liquid droppings like an eagle or sparrowhawk. A detailed, anatomical and behavioural comparison between South American cariamas and Secretary Birds is badly needed. Present-day Secre-

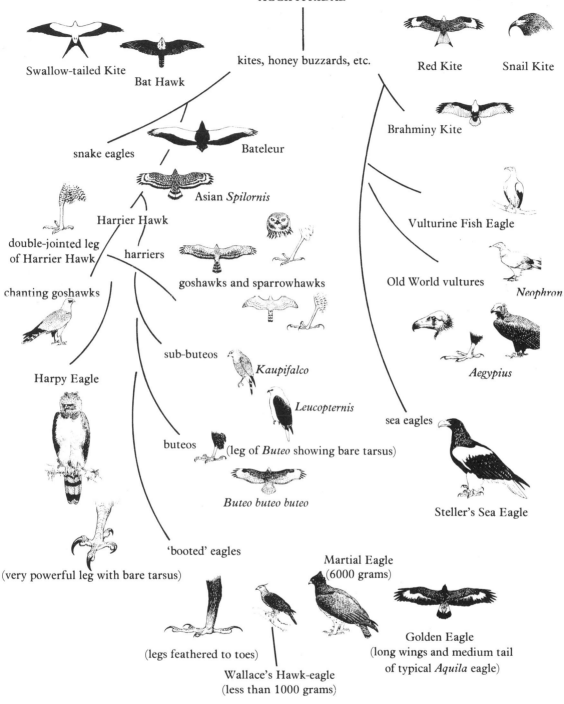

ACCIPITRIDAE

kites, honey buzzards, etc.

Swallow-tailed Kite

Bat Hawk

Red Kite

Snail Kite

Brahminy Kite

snake eagles

Bateleur

Asian *Spilornis*

Harrier Hawk

double-jointed leg
of Harrier Hawk

harriers

chanting goshawks

goshawks and sparrowhawks

Vulturine Fish Eagle

Old World vultures

Neophron

Harpy Eagle

sub-buteos

Kaupifalco

Leucopternis

Aegypius

sea eagles

buteos

(leg of *Buteo* showing bare tarsus)

Buteo buteo buteo

(very powerful leg with bare tarsus)

Steller's Sea Eagle

'booted' eagles

Martial Eagle
(6000 grams)

(legs feathered to toes)

Wallace's Hawk-eagle
(less than 1000 grams)

Golden Eagle
(long wings and medium tail
of typical *Aquila* eagle)

Fig. 3 Lines of evolution in the
family Accipitridae.

The Secretary Bird is a unique terrestrial species found in African grasslands; this immature has not grown its full long tail.

tary Birds are exclusively African, but some rather dubious fossils suggest that they were once more widespread, and the discovery of new fossils could necessitate adding another family to this superfamily.

3 The third suborder, the Falcones, includes only one family, the Falconidae. The Falconidae are distributed worldwide and include sixty species with thirty-seven in one large genus, *Falco*, comprising all the true falcons. All lay eggs with a reddish-buff inside shell; all moult their primaries in the same ascendant order; they share some anatomical characters; and the predatory falcons, at least, bob their heads up and down (like owls); many apparently do not vigorously squirt their droppings but allow them to fall below the perch. Old-time falconers noted this and called the droppings 'mutes', compared to the liquid accipitrine 'slice'. Falconidae resemble Accipitridae in sexual dimorphism, in having hooked beaks and talons, and even in some cases being carrion or snake eaters. Electrical analysis of their eggwhite proteins, however, indicates that Falconidae are not as closely allied to other birds of prey as they appear to be.

The Falconidae can again, for clarity, be divided into three subfamilies. There are Polyborinae, caracaras, milvagos and other aberrant species of the genera *Daptrius*, *Phalcoboenus*, *Polyborus*, and *Milvago*. These include some very curious examples of specialization; one *Daptrius*, for instance, eats wasps like the unrelated Old World Honey Buzzard; several eat carrion. Many of these aberrant falconids are rather large, and include the largest of the family. They make their own nests but their eggs resemble those of other Falconidae. They occur only in South and Central America. This also applies to the Herpetotherinae, including the Laughing Falcon, *Herpetotheres*, and the five extraordinary forest falcons of the genus *Micrastur*. No forest falcon's nest has ever been found but, like many true falcons, the Laughing Falcon breeds in hollow trees and makes no nest. It also looks more like a true falcon than a caracara, but eats snakes, partly filling the ecological niche of the Old World snake eagles.

Pygmy falcons and falconets are not obviously linked to the very odd Herpetotherinae; but small as they are, they are recognizably similar in shape to the true falcons of the genus *Falco*. The Spot-winged Falconet, *Spiziapteryx*, occurs only in South America, but is probably fairly closely related to the African Pygmy Falcon, and the eastern Fielden's Falconet. The true falconets, *Microhierax*, the smallest known birds of prey, are all oriental, but are clearly just tiny, blocky little insectivorous falcons breeding in old barbet and woodpecker holes.

The genus *Falco* includes more than half the family varying from small, sexually dimorphic, insectivorous kestrels, to large, powerful, bird-eating falcons such as Gyrs, Lanners, or Peregrines. This genus can be divided into sub-genera or superspecies groups but all are recognizably falcons. All pygmy falcons, falconets, and true falcons make no nests of their own but breed in other birds' nests, holes in trees, or ledges on cliffs; any old accounts of building stick nests are erroneous. All lay reddish eggs, except the pygmy falcons and falconets which lay white or whitish eggs in dark cavities. This evolutionary line perhaps culminates in the great, swift falcons, the Gyrs, Sakers, and Lanners; and finally in the Peregrine Falcon, with sixteen races which occur almost worldwide, and which was a hot candidate for the title of the world's most successful bird until pesticides struck it within the last thirty years.

This brief survey of the classification and distribution of falconiforms shows that there are many apparent gaps, anomalies, and points that cannot at present be explained, often because the species themselves are little known. Is the Osprey really related to primitive kites, or does it just not need a bony eyeshield because it does not have to crash into bushes and trees in search of its piscine prey? Is the vegetarian Vulturine Fish Eagle a vulture or a fish eagle, or something different altogether? Is the Secretary Bird a bird of prey at all, as it seems at present to be? Evidently, in answering these conundrums there is room for new anatomical and other laboratory work, but especially for often elementary observations on behaviour, which is little recorded even in some very common species.

Habitats and their inhabitants

In discussing birds of prey in relation to their habitat we can at once eliminate oceans because no bird of prey is typically oceanic. Several sea eagles and the Osprey may be commonest along sea coasts but they also occur inland. Others may sometimes be common along coastlines, notably when migrating or in winter, but are not strictly dependent on such habitat. Oddly, two unusual falcons, Eleanora's and the Sooty Falcon, may be as dependent on the sea as any diurnal raptor because they breed on small sea islands and catch birds migrating over the sea. No diurnal raptor, however, is truly dependent on the sea as distinct from the land where all must roost and breed.

On land, no birds of prey exist in ice-covered Antarctica so that we need only consider the Old World – Europe, Asia, Africa, Australia – the New World – North and South America – and islands in oceans. Most birds of prey either cannot cross or dislike crossing large bodies of open water. To reach and colonize islands, they must inevitably cross open water, so that even oceanic islands, which include a large variety of habitats, support few species of birds of prey.

In Hawaii (which has a great variety of habitats from sea-level to nearly 4200 metres) only a single buzzard species occurs. In more than 11 000 square kilometres of Viti Levu, Fiji with a variety of tropical forest and savanna habitats, only three birds of prey occur – an accipiter, a harrier, and the almost cosmopolitan Peregrine Falcon. More than 260 000 square kilometres of New Zealand, again with very varied subtropical and temperate habitats, supports only one falcon and a race of the same adaptable Marsh Harrier. Similar areas of varied habitats in tropical Africa or South America would support fifty to eighty species of residents and migrants. Thus, the number of raptor species occurring on islands apparently depends more on their remoteness than on the varied habitats occurring, though this may also be affected by the lesser variety of available food animals.

The accipiters inhabiting the island groups of Australasia and Oceania are more varied than anywhere else in the world. New Guinea supports eight resident species, New Britain five, the Solomon Islands four, Fiji one, and Samoa none. Raptors inhabiting such oceanic islands are usually enterprising members of widespread or successful types, harriers, accipiters, buteos, or falcons. There are no curious and unique survivals from the distant past. No flightless birds of prey, dodo-like, were exterminated by early mariners, though species such as the Mauritius Kestrel, now confined to small islands, are the most threatened. Some island species have 'primitive' plumage resembling immatures of mainland types, and a few, such as the Philippine Monkey-eating Eagle, may be relicts of once more widespread groups.

Neglecting islands, the major land masses of the world broadly contain nine main habitats, based chiefly on latitude and, therefore, on temperatures, from open, treeless Arctic tundra south through various types of woodland and steppe to equatorial tropical forest. The broad distribution of these habitat belts is modified by rainfall; most of the great deserts lie between subtropical woodlands and tropical savannas. Nor can we say that within certain latitudes a particular type of habitat will inevitably occur in convenient concentric circles because the major habitat belts are also modified by topography and by ranges of mountains which, because of their height, create temperature conditions not typical of the prevailing latitudes and which modify the expected vegetation.

Thus, the summit plateau of the Cairngorms in Scotland lies near the northern fringe of the temperate deciduous woodland zone but is covered with vegetation resembling Arctic tundra. The whole area is too small to support a population of typically Arctic raptors, but the Dotterel, Snow Bunting, and Ptarmigan, as well as several Arctic plants, can survive. On a bigger scale, along the Rocky Mountain chain and Californian Sierras, coniferous forests persist southwards into the subtropics and permit typically northern goshawks to breed south into Mexico. The Andean chain, likewise, enables some typically southern species to reach and even pass the equator, and supports one or two species of its own. The high tropical Ethiopian moorlands resemble Asian steppes abounding in rats, and are a favoured wintering ground of the Steppe Eagle.

Two special habitats supporting certain specialized raptors may occur almost anywhere from the Arctic to the equator. Most important is the aquatic habitat – seacoasts, large lakes, rivers, swamps, and marshes. Twenty-five species of diurnal raptors are found only or mainly in such aquatic habitats, and several of these are very widespread. The Osprey is almost cosmopolitan, and of all harriers, the Marsh Harrier, with its numerous races, is the most adaptable. Most aquatic raptors eat fish but a few eat crabs or molluscs. Harriers, which often hunt and breed in marshy ground, but also traverse open plains and steppes, form a link between mainly aquatic and dry-land raptors. Truly aquatic species cannot live without water and the animals it supports.

The aquatic habitat is mainly natural, increased or reduced by dams and swamp drainage. A wholly unnatural habitat of increasing importance is large, man-made towns and their environs which, surprisingly, support the highest known densities of any birds of prey. A few adaptable species have found that man provides nearly unlimited supplies of scraps and refuse and have quickly learned to live with him to their advantage. Most are scavengers such as vultures and kites and are more abundant in the tropics. A few adaptable, cliff-dwelling species, however, notably kestrels and even the magnificent Peregrine Falcon, have also colonized towns. No diurnal raptors are entirely dependent on towns, but some are undoubtedly much more numerous because of towns.

Often, there is no clear boundary between one major habitat and another; habitats simply merge by degrees. Thus, coming south from the Arctic Circle, conifers gradually give place to broadleaved trees of temperate deciduous woodland. Temperate or subtropical steppe and open plains gradually, with decreasing rainfall, become desert, and desert slowly responds to increasing rainfall to change to thornbush, acacia savanna, and broadleaved savanna. The most abrupt difference is between predominantly wooded areas and predominantly open areas; trees at once add a new vertical dimension to the habitat. Thus, the change between temperate woodland to open heathland is sharp, and that from tropical savanna to tropical forest is dramatic.

Often a rich diversity of habitat occurs in quite a small area. The flood plain of a big river running through forest creates patches of open grassland, pools, swamps, stonebeds, and so on. Human activity, through clearing and cultivation, may also create a richer diversity of habitat than in strictly natural areas. Or, a mountain range may rise suddenly out of a flat plain, bearing on its sides a whole range of life zones. Such local variations of habitat, due to drainage, human activity, or topography usually support a much greater variety of birds of prey than would occur in a tract of uniform habitat of similar extent.

Moreover, birds of prey are powerful fliers and every species could fly from one habitat to another, and many could soar to heights where they could not possibly hunt. The normally terrestrial Secretary Bird has been seen flying at 3600 metres and the normally aquatic Bald Eagle migrates across near-deserts in the western states of North America.

Naturally, therefore, many birds of prey can exploit more than one habitat. Even the most versatile species, however, prefer one type of vegetation but do almost equally well in others and can survive in yet more. Others like the edges of one habitat or another, typical of neither. The European Common Buzzard and the American Red-tailed Hawk may prefer a mixture of deciduous woodland and patches of open heath or grassland but they also do well on mountain moorlands, scrub-covered hills, or even (the Red-tailed Hawk) open desert with cactuses. Skulking sparrowhawks and goshawks which normally live in dense forest often emerge into the open to hunt. On the whole, it is less surprising that some birds of prey can exploit many habitats than that so many, especially forest species (some highly specialized), are almost confined to one main habitat and reluctant to leave it.

In the review of habitats and their inhabitants which follows, I have listed each raptor species first in the preferred or favoured habitat where it breeds. Where a species does well in several habitats, I have listed it for all of them, but where a raptor is mainly found in one habitat, and only occasionally in one or more others, I have listed it for only one. Thus, the American Broad-winged Hawk is typically an inhabitant of temperate, broadleaved woodland but local races occur also in subtropical and even tropical forests on West Indian islands. On the other hand, you could expect to see a Golden Eagle in any mountainous country from the Arctic to the Mediterranean, or from Alaska to Mexico.

Our ignorance of the precise preferences of many species hampers any accurate attempt at such a list. The habitat preferred by many South American, tropical Asian, and Australasian species is either not known or is imperfectly defined. Again, in one part of its range a bird of prey may prefer one habitat but may behave quite differently elsewhere. The Imperial Eagle perhaps prefers low-lying plains but in some areas nests on mountainsides. Sometimes, too, the only detailed study available may refer to a small fraction of the range, or to somewhat atypical habitat. Thus, the American Short-tailed Hawk has been studied only in Florida, right at the northern edge of its vast Central and South American range. We can only assume it behaves the same way elsewhere. My own protracted studies of the African Crowned Eagle have all been done in relatively open country with small forest patches; in the deep forests of Gabon I might find its habits and prey preferences quite different.

Subject to these reservations and provisos, the more varied the terrain and vegetation, the greater the variety of food supply, and consequently the greater the number of raptor species found. This is why Spain is such a good part of Europe for birds of prey. Treeless, freezing tundra and scorching, stony desert can support few species, whereas, warm subtropical, wooded hillsides, or tropical savannas support many. The harsher the conditions, the fewer the species of birds of prey (or others) that can survive. Broadly, this is due to limited variety, numbers, and habits of food supply. Other factors are also involved, however, and it does not inevitably follow that because a food supply exists a raptor will exploit it. Attempts to explain distribution, habits, and abundance on the basis of one factor alone run into trouble when applied to similar species or other habitats.

Do not attempt to digest all these details at once. Skim through and return to study particular habitats in detail as you choose. An understanding of the general relationships between habitat and way of life is essential for most of what follows.

Tundra

A huge belt of Arctic tundra circles the world's landmasses just south of the polar icecap. There is no true tundra in the southern hemisphere. Tundra lies almost wholly north of the Arctic Circle and covers vast tracts of Russia, Siberia, Alaska and Canada, and lesser areas of Norway, Sweden, and Finland. Typically, it is totally treeless, the tallest vegetation being lowly, creeping shrubs. The ground is permanently frozen hard, a few centimetres below the surface, in permafrost, and is largely snow covered in winter. In the short (three months) warm summers, with twenty-four hours of daylight, the ice melts, but the underlying permafrost prevents drainage, so that innumerable small pools and bogs develop on the surface. In the height of winter, twenty-four hours of daylight is replaced with twenty-four hours' almost total darkness, so that a diurnal bird of prey cannot even see to hunt, and the ground freezes hard again. In the brief summer, millions of migrant waders, ducks, geese, and small birds arrive and breed, voles and lemmings multiply, and there is a short season of great abundance. By August, most of the migrant birds have left, the young often preceding their parents, and by October the frost has clamped down. Thereafter, any permanent tundra resident must make do with scanty pickings.

All the diurnal raptors of the tundra feed mainly on living prey, though the Rough-legged Buzzard and Golden Eagle also eat some carrion. The most abundant large animals of the tundra, caribou and reindeer, are much too big, except as new-born calves, to be killed by any diurnal raptor, and in winter carcasses would freeze solid and be inedible. Thus, raptors must depend on the abundant live prey available in summer, or the small numbers of permanently resident species present in winter.

Only four diurnal raptors are typical of the tundra, the

Rough-legged Buzzard, Golden Eagle, Gyrfalcon, and Peregrine Falcon, and only the Gyrfalcon and the Rough-legged Buzzard are really characteristic of the tundra as preferred habitat. The Peregrine and Golden Eagle occur and breed in tundra but are more numerous elsewhere. Only the Gyrfalcon is a more or less permanent resident; the others migrate south from the tundra in winter, and even the northern tundra Gyrfalcons are partially migratory. All tundra residents are Holarctic, that is, they occur in tundra all round the pole, not only in Eurasia (Palearctic) or North America (Nearctic).

The Golden Eagle and the Rough-legged Buzzard are principally mammal eaters, the Eagle preferring ground squirrels and Arctic Hares while the Buzzard feeds on voles and lemmings. Both eat some birds but depend on mammals. The Peregrine and the Gyrfalcon, on the other hand, depend upon birds though the Gyrfalcon also eats some mammals, notably lemmings. In Alaska, 95 per cent of the Peregrine's food is birds but astonishingly one has been seen to wade into a cold river and catch a fish!

Gyrfalcons can stay in the tundra almost all year round because their favoured prey is Ptarmigan, also resident year round. The breeding success and population dynamics of the Gyr are tied to those of the Ptarmigan. If the Ptarmigan are plentiful in spring, Gyrfalcons breed and rear good broods, but if they are scarce the Gyrs may not breed at all, or rear smaller broods. The Peregrine and the Golden Eagle also eat Ptarmigan but the Peregrine takes more small birds, especially waders and passerines, than the Gyrfalcon. Thus, both can live in the same areas with comparatively little competition. If they compete, the Gyr wins.

The voles and lemmings preyed upon by Rough-legged Buzzards are exposed to view only in summer, but in winter, can live beneath the snow feeding on the buried, dormant, or dead vegetation. Thus, the Rough-legged Buzzard finds enough to eat in summer but is forced to migrate south in winter to areas where it can find rodent food.

Several mammal-eating, tundra species, such as the Rough-legged Buzzard, Snowy Owl, and skuas, show cyclic fluctuations of population in rhythm with their prey. In good vole years more Rough-legged Buzzards breed, lay bigger clutches, and rear larger broods. In a good year the population might more than double, but in a bad year increase only by about a quarter. Following good years, too-numerous young must migrate further and may have a poorer chance of surviving the winter further south. On balance, however, production of large broods in good years by Rough-legged Buzzards and other tundra raptors must lead to better survival for the species as a whole.

The tundra is the harshest of all environments for diurnal raptors, much harsher than deserts where the temperatures are generally high, even if food is scarce. Though only four species breed in the tundra, another three or four (some sea eagles, and even Swainson's Hawk) may cross stretches of tundra without having to depend on it. The main limiting factor is the long, hard, dark winter, reducing the more-or-less permanent inhabitants to the magnificent, snow-white Gyrfalcon, which can live here because even in winter it can catch enough Ptarmigan, themselves snow white and hard to see.

Taiga or temperate coniferous forest

As I said, trees immediately add another vertical dimension of habitat to life, and consequently enrich the environment. The first miserable, stunted conifers at the southern edge of the tundra are still trees. Further south they grow bigger to form, in serried ranks, the world's largest continuous tract of forest. This northern conifer forest occurs mainly between latitudes 50° and 65° north but naturally does not have a straight edge either in the north or in the south. In the centre of North America and in Siberia the continental climate pushes it south to Lake Superior and Lake Baikal. No taiga occurs in the southern hemisphere. The huge tracts of dense conifer forest are broken by patches of moorland and bog (in America called muskeg) where raptors that cannot catch prey in thick cover can live. Here the Peregrine Falcon so far departs from its normal, cliff-loving ways (in Europe, not in Canada, Alaska, or Siberia) that it breeds on the ground on dry hummocks.

Three times as many raptors inhabit this coniferous belt as the tundra, and are rather more inclined to be permanent residents, or partly migratory. The Hen Harrier and the Pied Harrier breed here in openings or clearings at least at the southern fringe of this zone. The Rough-legged Buzzard, as characteristic of taiga as of tundra, is joined by northern, migratory races of the Common Buzzard, and the Red-tailed Hawk. Three typical, short-winged hawks, the Goshawk, Sparrowhawk, and Sharp-shinned Hawk occur here, and the Goshawk and Sharp-shin prefer this habitat. The Golden Eagle is not really typical of such dense woodland but breeds sparingly on cliffs where there is enough open country in which to hunt. The Gyrfalcon, in this zone much less likely to migrate, and the Peregrine are also characteristic and are joined by the diminutive Merlin, which here usually breeds in old birds' nests in trees.

Thus, the taiga supports at least twelve raptors, of which all but one, the Golden Eagle, are quite typical.

Eight are migratory or mainly migratory, while many individuals of the other four especially from the northern limits of the taiga must also migrate. The increase in the number of species from four in tundra to twelve in taiga is primarily due to the presence of trees, enriching the environment, creating a wider variety of habitat, and supporting more and more varied prey.

Migration is still forced on most of the taiga raptors in winter because the snow lies much deeper among the trees than in the open tundra, where the wind can blow it off whole hillsides. The species still mainly resident are bird eaters: the Gyrfalcon, which is here barely migratory; the Goshawk, big enough to kill woodland grouse, hares, or squirrels; and to a much lesser extent, the Sparrowhawk, which can catch a few small birds. As in the tundra, the three buzzard species mainly dependent on small mammals, which in winter retreat beneath the snow or hibernate, must migrate, but the few Golden Eagles that live in the taiga can probably survive year round in their home ranges on such animals as hares with some carrion. The Peregrine, Merlin, and both harrier species migrate south. The Pied Harrier migrates much further south than does the Hen Harrier, which scarcely reaches even the subtropics.

All tundra raptors are big or fairly big, the smallest, the male Peregrine, still weighing 600 grams or more. In the taiga, five species weigh less than 500 grams, and the Sparrowhawk, Sharp-shinned Hawk, and Merlin weigh less than 300 grams. The smaller species are all migratory, the larger more likely to stay in the habitat. Small species require more food per unit of bodyweight to stay alive and are less able than larger species to survive the rigours of a taiga winter.

Although the taiga is still too severe an environment to support human crops and cultivation (except locally along the southern fringe) it is markedly less severe than the tundra. The summer is longer, the winter shorter, and the hours of total darkness fewer, so that diurnal birds of prey have a better chance to hunt to live, even in winter.

Temperate deciduous woodlands

Taiga gradually gives place to this type of woodland further south. On the southern fringe the dark, evergreen conifers mingle with patches of birch and aspen which, in autumn, make a glorious golden contrast against the sombre green of the firs and spruce. These deciduous trees drop their leaves in winter in response to cold. The first, sharp, autumn frost kills the leaves and the winter gales blow them off, so that the trees appear lifeless and naked until the increasing warmth of the following spring bursts the dormant buds and the forest is green again.

The most familiar of all the world's habitats to people in Europe or North America because it is the one that most of us inhabit, this is also the habitat most affected – in Europe for many centuries – by human activity and development. Few considerable tracts of this woodland or forest, either in North America or in Europe, have not been more or less severely altered by man, though in America it was almost virgin only 300 to 400 years ago.

This is also the first of the major habitats which occurs in both northern and southern hemispheres though the area in the south is small. In Patagonia and southern Chile, a tract of southern beech forest (*Nothofagus*), wet, cold, and gloomy, runs northwards along the Andes. Two species, a caracara and a buteo, seem characteristic of this tract. Three Australian accipiters and the Wedge-tailed Eagle occur in Tasmania and may be characteristic of small areas of this zone, though all are commoner elsewhere. Most raptors inhabiting deciduous woodlands (eighteen of twenty-three species), however, are northern because the greatest tracts of this woodland are in the northern hemisphere. Of these, only the Goshawk occurs in both the Old and New Worlds; eleven are Palearctic and six Nearctic.

Although many species living in this woodland type are wholly or partly migratory, thirteen of twenty-three or 56 per cent of the raptors are normally permanently resident. Some species which in the taiga are largely migratory, such as the Red-tailed Hawk, Common Buzzard, and Goshawk, here become permanent residents. Other species which first appear here, such as the Red Kite, Honey Buzzard, Lesser and Greater Spotted Eagles, the Hobby, and the American Broad-winged Hawk are wholly migratory.

Apparently, as the climate grows warmer, fewer species must migrate but those for which a particular habitat is the northern or southern fringe of their range must normally move to warmer climates in winter. This does not apply to all species known in this zone, however, because the essentially tropical Mountain Hawk-eagle apparently resides permanently in northern Japanese coniferous forest. Possibly, this is because it kills large and permanently resident prey, but very little is recorded on its habits here. The three south Australian accipiters are also resident.

This zone is all woodland or forest, with relatively little open space, so that most of the raptors which live in it are woodland birds. The most characteristic are goshawks and sparrowhawks (accipiters) and buzzards or buteos. Of the twenty-three species, eight are goshawks or sparrowhawks, and six are buzzards or buteonines. The three eagles that occur are all medium-sized or large woodland or forest species, and the Red Kite prefers woodlands. The wasp-eating Honey Buzzard breeds in deep woods, and the only falcon typical of this habitat, the Hobby, nests in other birds' nests in trees but feeds on insects and birds caught in the open.

In this zone, too, there is for the first time a marked separation of similar species between continents. Of eight goshawks and sparrowhawks, only one is Holarctic, two are purely Palearctic, two are Nearctic (North American only), and three are Australian. Among buzzards and buteonines, three are North American, one South American, and two Palearctic. Thus, although there appears to be a proliferation of similar, possibly competitive woodland species, in practice, the number that could compete in any locality is reduced to three or less by geographical isolation. Moreover, potential competition is further reduced especially in winter by the migratory habits of some species. The greatest variety is found in summer when food is likely to be most abundant.

Man has affected this habitat perhaps more than any other. In Britain, France, and other European countries most of the woodland has been converted during several thousand years, to open, cultivated fields. Small patches, often artificially planted, remain though in Europe are often much larger than in Britain. In North America the early settlers destroyed much of the woodland so that they themselves could cultivate and survive. More recently, much of this poor farming land has reverted to woodland though this is not quite the same as the virgin forests. The only reason the *Nothofagus* woods of South America survive human destructiveness is the inhospitable climate of Patagonia.

Thus, in the tundra or taiga the population of birds of prey is essentially little affected by man's activities, but in the temperate deciduous woodland the relative abundance of species has been greatly affected by man. How much, we do not know, because there are no accurate old population estimates.

Birds of prey have often been reduced by direct persecution, especially in Britain and Europe, but some also have benefited by clearing for cultivation. Probably, the Common Buzzard and the American Red-tailed Hawk thrive better in a mixture of open fields, rough grazing, and woodlands than in unbroken woodland, though in eastern England the Buzzard is prevented from reoccupying perfectly suitable country by direct persecution. American and European Kestrels (not characteristic woodland species) now breed regularly in patches of woodland hunting in the open fields. The Kestrel must now be much more abundant because of man's activities than in prehistoric times. On the other hand, the two European spotted eagles have certainly been reduced by man and in Britain the once abundant Red Kite was almost rendered extinct by persecution combined with improved sanitation.

Although no species has apparently been rendered totally extinct by man in this temperate woodland, some species have become rarer and others may have benefited. The population of birds of prey has adapted to the more or less gradual changes brought about by man's activities. Such changes are still going on. Some reduction in persecution has enabled the Buzzard to reoccupy large tracts of England, Wales, and Scotland since 1914, and the Goshawk disappeared as a breeding species after 1800 until recently when it re-established itself in the very large conifer plantations of the Forestry Commission established since World War II.

Thus, the temperate deciduous woodland is a habitat of special interest, not only in itself, but also as an example of how birds of prey can adapt to ecological changes brought about by man, and to human persecution itself.

Temperate moorlands, mountains, and marshes

Although no purist would accept any such loosely defined habitat, I feel I must include it because large tracts of such open country do exist, composed of open heath or grassland, with few small trees. Such habitat has no very extreme winter temperatures nor permafrost, and is often wet or boggy, so that it fits neither tundra nor the concept of semi-arid steppes and plains. Moreover, it is inhabited by a quite different group of birds of prey to the temperate woodland or coniferous taiga in the same latitudes and must at least be regarded as an extensive subhabitat occurring alongside but distinct from these zones.

Most of this habitat is northern but the extensive tussock-grass habitats of South America, as treeless as tundra, and of New Zealand and Australia are included; like northern types, they are often moist and boggy though their botanical composition is very different. They support certain very characteristic birds of prey which must be included somewhere. In Australia and Tasmania such moorland occurs mainly on the higher mountains but strikingly similar country runs northwards right along the Andean Spine to Peru, and occurs on the mountains of East Africa near the equator above 3900 metres. I have grouped these tropical Andean and African moorlands, which have nightly frosts but no winter, in a special small habitat, tropical montane moorlands, discussed later.

Temperate moorlands, mountains, and marshes support twenty characteristic or typical breeding birds of prey and many of those from tundra, taiga, or even deciduous woodland occur in it on migration. The converse is not normally true, because although many woodland species emerge into the open to migrate or hunt, typically moorland species such as harriers will not readily enter woodland, though they may pass over small areas of it. They generally prefer to thread their way through or past forests by following rivers, fields or coastlines.

The characteristic species of moorlands and mountains are very different to those of temperate or coniferous woodlands. They include five harriers (two of which breed in open patches in the southern taiga) and six falcons, several of which are not exclusive to this habitat. One, Kleinschmidts's Falcon, *Falco kreyenborgi*, is a very rare species and its relationships are not clear but the known specimens came from southern Chile. The European Common Buzzard adapts well to this zone and the South American Red-backed Buzzard lives only in such open moorland country but extends north along the Andes into the tropics. The Andean Condor is as typical of this zone as of the Andes further north. Finally, four omnivorous South American caracaras and milvagos occur, generalized scavenger-predators which here take over the role of large crows on northern moorlands; there are no crows in Patagonia. Their habits explain how quite a small tract of open country in southern South America can support eight out of the twenty species typical of this habitat.

Of twenty species, seven (35 per cent) are migratory, but of these only four (20 per cent) are wholly or mainly migratory. Some, such as the New Zealand race of the Marsh Harrier, are resident but the Tasmanian Marsh Harrier migrates. Perhaps the Marsh Harrier should not be included here at all because it is more characteristic of marginal aquatic habitats. In New Zealand and Australia, however, it may nest in marshy places but hunts readily over open moorland. The proportion of full migrants to resident species is much lower than in temperate deciduous woodlands, reflecting both the comparatively mild temperatures and also, perhaps, the relative ease of finding and catching prey in open country.

This type of country is also much affected by man's activities. As in Britain, moorland may often have developed or extended and now be maintained as a result of destruction of former woodlands followed by burning and grazing by sheep and deer. Game preserving and sheep rearing affect the status of some species. The Golden Eagle has been, and still is, persecuted by sheep farmers, and it and several other species have been severely persecuted in Britain by proprietors of grouse moors. In South America the introduction of sheep may have led to persecution of caracaras and milvagos, but doubtless also provided quantities of carrion not there before, enabling these birds to multiply greatly. Such assessments are informed guesses in the absence of any reliable population data.

This habitat is still being largely managed and manipulated by man. It is burned at intervals for sheep raising and in places is also being afforested. In Britain, large tracts of such open grass or heather moorland have been planted by the Forestry Commission in recent years usually with Alaskan Spruce; in effect, such moorland is being reconverted to taiga. It seems quite clear that these plantings encourage woodland accipiters, such as the Sparrowhawk and Goshawk, and discourage open-country species such as the Golden Eagle. When young planted trees are five to ten years old, these plantings become very attractive to nesting Montagu's and Hen Harriers and have been a prime factor in the increase of the latter. Once the trees grow up, the harriers disappear. The young plantations probably resemble a natural, semi-open tract of scrubby, rank vegetation rich in small mammals and birds rather like the natural breeding haunts of the Pied Harrier in Siberia and the Hen Harrier in the openings in coniferous forest zones of North America.

Probably, the birds of prey in this habitat have been more intensively studied than in any other except temperate deciduous woodland because the areas concerned are accessible to energetic European and American ornithologists, and because it is usually much easier to study birds of prey in open country. Trees add a new vertical dimension but they also obscure the view and make large, wide-ranging raptors hard to follow. South America, where neither the resident Cinereous Harrier nor the abundant caracaras and milvagos have ever been studied in any depth, is an exception to the generally high standard of knowledge. The Harrier would probably behave much like a Hen Harrier but the caracaras are like no other birds of prey and deserve the full study that could so easily be carried out.

Subtropical semi-arid woodlands and mountains

This zone is the first of the world's really rich habitats for birds of prey. It supports no less than seventy-three species of which sixty-five are quite typical, though some are also typical of other habitats. Relatively warm temperatures and comparatively light snowfall reduce the likelihood of a foodless winter while even in mid-winter there are some hours of good, bright daylight permitting hunting. Diversity of topography and ecology, and consequently of food supply, all make for a wider variety than can exist either in temperate areas to the north or deserts to the south. Broadly, this zone is characterized by a hot, dry summer and a mild, wet winter and spring. Such climates are typical of the whole Mediterranean basin, parts of the Middle East,

California, northern Mexico, Argentina, Chile, South Africa, and Australia. In the southern hemisphere, of course, the seasons are reversed, winter occurring in the northern summer, a fact important for some migrant raptors which commute between the two hemispheres.

Most growth occurs in the short spring between the cool winter and the summer drought, so that tall forests are not typical of this zone. Most trees are xerophyllous, that is, they have hard, shiny or waxy, evergreen leaves resistant to drought. There are pine forests, too, but these have the open, branching habits of *sapins* in southern France or Spain, or in coastal California rather than the dark, closed spruce and firs of the taiga. In Australia, eucalypts replace all other trees, and have been extensively planted elsewhere in this zone. The enormous Californian sequoias, dwarfing even 60 metre pines growing on the same sierras, are doubtless the result of locally higher rainfall but the evergreen oaks of the San Joaquin and Sacramento Valleys of California are strikingly like those of Spain or Morocco.

Where there are no trees, the country is often covered with dense mixed thicket, known as *maquis*, chaparral, *fynbos*, and so on. Its composition differs but both in South Africa and the Mediterranean, a main component is Giant Heath, *Erica arborea*, or *bruyère*, the stuff of 'briar' pipes. Left alone, such thickets soon become impenetrable, and no bird of prey could catch anything in them. They are usually highly inflammable in summer, however, and man or lightning sets them alight. Mountain slopes, through burning and overgrazing, then often become bare and stony, with fangs of cliffs protruding but trees and woods are usually also associated with lower mountain valleys. In spring, slopes and lowlands alike are covered with a magnificent profusion of beautiful wild flowers, whether in Crete, California, Australia, or the Cape Province.

Climatically and scenically, these are among the most delightful countries in the world, never really cold, with plenty of sunshine and where life without the gentle, beneficent, civilizing influence of abundant fruit and wine is almost unthinkable. The relatively fragile, semi-arid environment, however, will not survive the same amount of misuse as, say, a virtually indestructible peat bog in Scotland. The original soil and vegetation can be destroyed and lost by careless farmers in very few years. More than 2000 years of continuous misuse have severely damaged natural resources all round the Mediterranean. Though such damage may be less evident elsewhere, the pattern is the same, of numbers of people attracted to these areas because of their pleasant climate, tearing down the forests to cultivate, and when that fails, setting the *maquis* or chaparral alight to graze their goats, by then all that will thrive. In historical time it does not take long to ruin this environment and some parts, notably southern France and California, have already become hideous conurbations.

Such facts naturally affect birds of prey in these areas but although many species have become rarer, probably none has actually become extinct. The nearest to extinction is the California Condor, reduced to about forty and perhaps unable in the long run to withstand the intense pressures of human population and development that beset it. Some other species may, on balance, have benefited by the human destruction of the environment. It is a pity that Aristotle, who certainly observed that eagles were territorial, could not have been a bit more precise about their population density in his day.

The species truly characteristic of this zone cannot always be clearly known because, particularly in Australia and South America, habitat preferences may not be precisely defined. Only tropical forests and savannas, however, have more than the seventy-three species occurring here. In addition to the breeding species, the northern parts of this zone receive many of the migrants from temperate woodlands and some go no further south. It is as if they journeyed as far as the south of France or Spain and then said, as you or I might, 'This will do nicely, thank you'!

Although the Andean Condor has already appeared in Patagonia, and although the American Turkey Vulture breeds in woodlands as far north as southern Canada, vultures, especially Old World vultures, first become characteristic of the raptor population in the subtropics. Old and New World vultures are not closely related but both feed on carrion and they have long nesting cycles. The availability of carrion cannot be the only, even perhaps the main, reason for their greatly increased abundance here, because there is probably less carrion available in the relatively mild winters of the subtropics than in colder winter climates. If the supply of carrion is related to the time of maximum regular demand, however, when the parents are feeding young in the nest, it may be that some carrion available in summer because of increasing drought is more important than superabundant carrion available mainly in winter and spring – and which would rot anyway in the warm subtropics.

Also, in terms of overall population dynamics, it may be more important for vultures to have some carrion available late in summer, when the young must compete for it with stronger, older adults, than abundant carrion in February and March when adults, which can fast for weeks at need, are incubating eggs.

The lengthy incubation and fledging periods of vultures may also prevent breeding in colder climates. Incubation lasts forty-five to fifty days or more and fledging ninety to 110 days. Thus, the entire breeding cycle cannot be completed in less than five months, more likely six. Vultures need a long summer to breed and rear young, and could not breed at all in Alaska or even Perthshire. The Andean Condor, with a very long breeding cycle but breeding in Patagonia, appears to contradict this theory; but the winters in Patagonia are said to be less severe than in comparable temperate latitudes.

Some vultures migrate in winter, partly or wholly. Again, the Egyptian Vulture, the smallest, is the most strongly migratory but the unrelated American Turkey Vulture is also migratory from its northern range and some European Griffons migrate south. On the other hand, the biggest and most impressive species, the Lammergeier, European Black Vulture, and Andean and California Condors are all mainly or wholly resident in their home ranges year round. Again, the smaller the species, the more inclined they are to migrate. Vultures are, in general, much more abundant in the tropics than in this zone but this is at least partly due to improved sanitation and veterinary services in the last 200 years.

Snake eagles, which are typically tropical, also appear first in the subtropics, because they can catch enough lizards and snakes to breed and rear young. Snakes and lizards must have warmth. They decrease in numbers, and in variety of species, north of the subtropics, and are sluggish and inactive for much of the year. The European Snake Eagle, or Short-toed Eagle, is wholly migratory in this zone, though its tropical African races are sedentary. It arrives about March when the lizards and snakes emerge, breeds, and departs southwards before the first chill of autumn makes the reptiles retreat underground. Thus, the breeding season must be compressed into the period when food is abundantly available. Likewise, Honey Buzzards in temperate woodlands must also compress their breeding season into the summer months when wasps are abundant; young Honey Buzzards migrate almost as soon as they can fly. Honey Buzzards, like other northern migrants, pass through the subtropics, but do not commonly breed, perhaps because the summers are too dry for abundant wasps.

Among other birds of prey, the number and variety of species in groups already familiar increase. Accipiters, for instance, increase from eight regulars in temperate woodlands to twelve, including the same three wide-ranging Australian species. Two of the four new accipiters, the Levant Sparrowhawk, and Grey Frog Hawk are migrants; both are small. Buzzards and their near allies increase from six to fourteen including several species already mentioned and some new ones characteristic of the subtropics.

The variety of eagles, never more than and usually less than three in any zone further north, increases in the subtropics to nine, eight of them characteristic. They include

Above Two White-backed Vultures; belonging to the Old World vultures (Accipitridae), they are not closely related to New World vultures but have bare necks, an example of parallel evolution.

Right A Common Caracara; an aberrant new world falconid, behaving more like a vulture than a falcon, making its own nest, but moulting in the same way as a typical falcon.

Above The boundary between tundra and taiga is typified by large stretches of open ground with small trees colonizing suitable tracts, glaciated rocks, and snow on high ground.

Left The Gyrfalcon, the world's most powerful falcon, is the only permanent raptorial resident of tundra, living largely on Ptarmigan.

several big, wide-ranging species such as the Golden Eagle, Wedge-tailed Eagle, and Verreaux's Eagle. The so-called hawk-eagles of the genus *Hieraaetus* first appear here and one, again the smallest, the Booted Eagle, is a breeding migrant. It has recently been found breeding not only in the Mediterranean subtropics but in South Africa as well, like a few other birds, including the European White and Black Storks and European Bee-eater, which breed in both northern and southern hemispheres.

Hawk-eagles (still more varied in the tropics) do not normally eat carrion but kill all their own prey. Perhaps this suggests that, in order to live at all in colder climates, eagles must either migrate completely (as do the spotted eagles) or be able to survive on carrion for part of the year. Also, the bigger the eagle, the easier it can tear its way into a part-thawed carcass and the less food it needs in relation to bodyweight, so that a big, sluggish eagle has a better chance of surviving cold winters than a small, active one.

Falcons increase dramatically in this zone from five in temperate moorlands to seventeen, some also characteristic of tropical zones. All temperate species except the Merlin occur here. New ones include two wholly migratory small species, the Red-footed Falcon and the Lesser Kestrel, both more inclined to hunt in steppes or open plains, but dependent respectively on old crows' nests in trees and cliffs to breed so included here. In Europe and Asia, races of the Saker and Lanner occur in subtropical woodlands but are more characteristic of subtropical steppes and tropical savannas respectively. The exact habitat preferences of some large Australian falcons are unclear but because they do not seem to be confined to open country and breed in trees I include them here, with some reservations.

The limitations on small size apparently imposed by cold in the zones near the poles here virtually disappear. The subtropical woodlands support some of the largest raptors, especially vultures, and some very small kestrels and sparrowhawks. Even within any well-represented group, for instance, among the eleven New and Old World vultures, twelve accipiters, twelve buteos, and seventeen falcons, the size range found in the subtropics is almost the same as for the whole group or genus. Low temperatures in fact, cease to be a true limiting factor on life. The raptor fauna of this zone demonstrates clearly how a warmer climate results in a much greater diversity of species, while there are obviously more individuals of some species than in temperate zones. Although the subtropics cover a much smaller area than the temperate woodlands or taiga, the increase in the number of species found is relatively huge. Migration is also reduced and only a quarter of all the species do or must migrate.

Much research is still needed, especially in the southern hemisphere, in Australia and America, before the ecological requirements of many subtropical species are precisely understood. In Europe and southern North America, however, the needs and behaviour of most species are quite well understood, as a result of advances in knowledge in the last ten years.

Temperate to subtropical open steppe, plains, or prairie (steppes)

These habitats occupy vast tracts of North and South America, central Europe and Asia, Australia, and parts of South Africa. Again, I have over-simplified a variety of different ecological types with roughly the same physical characters. They are flat or gently undulating, with no mountains or big cliffs. They are covered with short or long grass, low shrubs, or bushes; the few trees are found along watercourses or are planted and tended by man. Being flat and treeless, they are often windswept, very hot in summer, and cold in winter. Rainfall is low, always less than 750 millimetres, more often 250 to 500 millimetres per annum. Water is often scarce, sometimes brackish, and nowadays very often behind man-made dams.

The climatic factors prevailing dictate the type of land use. Before human exploitation, such areas were the home of vast, nomadic herds of hoofed mammals such as bison or gazelles but these have been largely or wholly exterminated. The people who tried to cultivate these areas, in America or Russia (because they looked easy), were often repaid by the creation of a dust-bowl in which they starved. The steppes and plains generally cannot be cultivated except under irrigation and, since the native mammals have been exterminated, are largely used for stock raising. This, under good management, changes the habitat far less than cropping which replaces a wide ecological variety with a monoculture of an edible crop. Good management is, alas, the exception rather than the rule but even under poor management the effect on birds of prey may not be very severe. Poor management also causes heavy stock loss producing abundant carrion, and the invasion of unpalatable bush and shrubs, which may actually improve the ecological diversity and the variety of food supply.

Often, tracts of steppe and subtropical wooded mountains intermingle, as for instance, all along the Rocky Mountain chain, the foothills of the southern Andes, or the intermontane basins of the Atlas in north Africa. On the Great Plains of Colorado you may see Golden Eagles sitting on windmills, just as you might see a Steppe Eagle in southern Russia. Swainson's Hawk and the Ferruginous Hawk, typical North American plains' species, may nest

Right In autumn the colours of dead leaves in temperate deciduous woodland are brilliant, but herald a period of scarcity when most raptors of this habitat may migrate.

Far right In winter snows mammal-eating species such as Buzzards cannot easily find food, but bird killers such as this healthy Sparrowhawk may find their prey weakened and easy to catch.

Below Maquis. In spring the mountains of Crete are still snow capped but the flowers are in full glory.

Below right A pair of Bonelli's Eagles at the nest. The rocky cliffs provide nesting habitat and the open ground provides rabbits and gamebirds as food for this powerful, swift species.

within a kilometre of woodland Red-tailed Hawks. The Prairie Falcon will breed on any available cliff in open country, often a gorge, but as often must resort to wooded hillsides to find a breeding cliff.

Woodland or mountain species often move into adjacent areas of steppes and plains, especially to winter; relatively few exclusively inhabit one or the other. Large, cliff-nesting vultures such as griffons or condors would starve if they did not forage freely on the plains. Smaller species, such as Egyptian and American Turkey Vultures can breed in gullies in steppes because they need no big cliffs. Generally, where several closely related species occur in the same geographical area they are ecologically separated, that is, they make use of different habitats, but sometimes one of such species takes over altogether. In eastern Europe the Long-legged Buzzard occurs in subdesert steppes and is ecologically separated from the woodland Common Buzzard. The north-west African race of the Long-legged Buzzard, however, inhabits both plains and woodlands alike, behaving much like a Spanish Common Buzzard.

The abrupt changes of topography, rainfall, and drainage between mountains and steppes mean that many birds of prey can be in totally different terrain and vegetation within half an hour's flight. There is no such abrupt physical difference between steppes and subdesert or desert; one just merges gradually into the other as rainfall decreases. For these reasons, it is hard to decide which are characteristic steppe species, because so many also occur in other habitats. I have included fifty-nine species occurring and breeding of which forty-seven prefer steppes though many also extend into other habitats. A few new types or groups are apparently best adapted to life in grass plains or steppes.

Black-shouldered kites of the genus *Elanus* are typical plains' raptors. Three species occur in America; Europe, Asia, and Africa; and in Australia respectively, while the Letter-winged Kite occurs in drier central Australia. These kites all also occur in tropical savannas, and sometimes in temperate wooded grasslands but are typical of grass plains and steppes. All are small, grey and black, rat- and insect-eating birds which hunt by hovering like kestrels, and are at least partly nomadic or migratory. The Mississippi Kite is typical of North American grass plains and prairies, breeding in trees and consequently benefiting from plantings by man. It is completely migratory in most of its range. Another newcomer is the unique, terrestrial Secretary Bird, as typical of subtropical African plains as of tropical savannas, eating rats, large insects, and snakes in about that order of importance. It seldom flies though can fly quite well.

Harriers, predictably, are well represented with eight species of which the Pallid, Cinereous, Long-winged, and South African Black Harriers are more typical of this habitat than any other. The Pallid Harrier is almost completely migratory from Europe and Asia to Africa and India in winter but the others are either resident or partly migratory. The movements of the South American species are poorly understood and the Australian race of the Marsh Harrier apparently prefers grass plains and big fields to marshes.

Woodland accipiters are practically absent from the steppes, but the chanting goshawks, *Melierax*, typical of tropical savannas, do occur here. They hunt much on the ground, like miniature Secretary Birds and emerge into the open much more than true accipiters. Eight buteos or buteonines are typical of steppe; one, Swainson's Hawk, is so specialized for this habitat that it migrates almost completely from North American prairies to winter in South American pampas. Others are resident, partly migratory, or nomadic, including the largest and finest of all buzzards, the Ferruginous Hawk, Long-legged Buzzard, and the Asian Upland Buzzard, as big as a medium-sized eagle. The Bay-winged Hawk and White-tailed Hawk, and even the adaptable Red-tailed Hawk, may breed often in steppes though are less typical of this habitat.

The wholly migratory Steppe Eagle, now considered a race of the adaptable and widespread Tawny Eagle, was so well adapted to European steppes that it bred on the ground. Recent land use changes have forced it to nest in trees. Like many other steppe birds, it typically eats rodents. Other large eagles, including the Imperial and Australian Wedge-tailed Eagles are less confined to this habitat. The Wedge-tailed Eagle is widespread from temperate woodland to tropical savannas, and partly fulfils the carrion-eating function of non-existent vultures in Australia. No New or Old World vulture is exclusive to steppes and breeding vultures are mainly small. The European Snake Eagle occurs if it can find a few small trees for breeding, and is migratory.

Fifteen species of falcon, chiefly adaptable species not confined to this habitat, breed in steppes. In Turkestan, a small, pale-coloured race of the Merlin reappears. It parallels the desert and steppe Peregrine, or Shahin, which some regard as a separate species, *Falco pelegrinoides*, but which is more likely to be just a small, pale-coloured Peregrine. In Australia, falcons found in steppe also occur in other types of country. The falcons most characteristic of steppes and plains are the Eurasian Saker, the North American Prairie Falcon, the south and north African Lanner, and perhaps the Aplomado Falcon, though its habits are little described. The Lanner is much more

Harriers are characteristic birds of prairie and steppe; when only one species occurs it occupies all types of open country. This is the Australian race of the Marsh Harrier.

widespread and adaptable than the Saker which typically breeds in steppes and hunts rodents, hovering over the plains like a huge kestrel. In winter, the Saker migrates to become the powerful, swift, bird killer prized by falconers. On high Ethiopian moors swarming with rats, which superficially resemble steppes, the Lanner feeds on rodents just like the Saker.

Several kestrel species are also typical of steppes and in these areas are highly nomadic. A tendency to nomadism is perhaps the most noteworthy feature of raptors of the steppe. It is distinct from migration, as practised in the same environment by, for instance, Swainson's Hawk, in that much of the population may shift in the breeding season from one area to another, concentrating in areas with a temporary abundance of food. In steppes, as in the Arctic, birds of prey are often dependent on rodents. Characteristically, rodent populations pass through cycles of abundance, increasing from few to almost incredible squeaking numbers, then disappearing as the population crashes. Such a cycle may occur through spring and summer in one area and not in another. Evidently, it would then be advantageous for rodent-eating raptors to concentrate in an area of abundance, breed while the food supply lasts, and seek another area in which to feed and rear their broods the following year.

All four *Elanus* kites are nomadic in this way. The kites often roost gregariously, and may also breed gregariously; they are clearly more abundant during rodent plagues, though good studies are lacking. The Secretary Bird is also nomadic, though no-one knows whether it walks or flies from one place to another; probably it flies, perhaps at a great height. All kestrel types and the Australian Brown Falcon are partially or largely nomadic and the movements of the little-known Australian Black Falcon may be related to those of quail – themselves nomadic.

The familiar Common Kestrel, a permanently resident, apparently solitary nester in southern England, breeds colonially in parts of the steppe. This is not really surprising because in the polders of Oost Flevoland in Holland the population of the Kestrel, breeding in superabundant nest-boxes, fluctuated widely according to the food supply. These Kestrels defended less than half a hectare of territory round the nest-box actually used, behaving like colonial nomads.

Such almost non-territorial habits and a readiness to use one or other part of a vast tract of country according to available food supply would evidently enable steppe Kestrels to breed more successfully than if they were strongly territorial and evenly spaced well apart. The Saker, too, will apparently tolerate other Sakers breeding within a kilometre and may be less strongly territorial than most big, fierce falcons. Such adaptations, however, would not necessarily benefit the bird-eating Peregrine and Merlin.

Apart from ground-breeding harriers and the Merlin, the numbers of raptors breeding in the steppes may be limited by available nesting sites as well as food supply. Often, falcons or buteos are forced to use any little available cliff or earth bank and several different species may breed close together. The lack of large trees or bushes also limits breeding, and in recent years the population of several American raptors, notably Swainson's Hawk and the American Kestrel, has increased through the plantings of cottonwood trees surviving around dams or near, now often deserted, homesteads.

Of fifty-nine characteristic or occurring steppe species, only eleven are confined to steppe and grassland habitats; the rest all occur more or less widely in adjacent wooded mountains or tropical savannas. Only eight, about a seventh of the total, are truly migratory, but in another six species, many individuals are resident while others, perhaps mainly immatures, migrate. At least eight species are more or less resident but nomadic and probably some of the migrants are also partially nomadic with more individuals breeding in one area than another in different years. All the regular migrants breed in the northern hemisphere; the species of South America, South Africa, or Australia are nomadic rather than truly migratory. Too little is recorded of their movements, however.

One species, the Peregrine, is cosmopolitan, and one, the Merlin, Holarctic. Seventeen are primarily European or Asian but several breed also in Africa. Nine are primarily North American, ten primarily South American, ten African, and twelve Australian. The number of species typical of steppes is less than twenty on any continent but species from other habitats often occur in steppes, especially in winter. The North American Great Plains support large numbers of Rough-legged Buzzards and Golden Eagles in winter, although neither the Buzzard nor the Eagle breeds in this habitat.

Although steppes and plains have a less rich raptor fauna than subtropical wooded areas, it is still much richer than temperate moorlands, again showing the beneficial effects of a milder climate. The special interest of the steppe raptors is the tendency to nomadism, and colonial breeding in species which may be solitary elsewhere. This is a rather different method of adjusting to fluctuating food supply than the increased clutch and brood size typical of Arctic raptors. Much more work is needed on this and because nomadism is specially characteristic of Australian raptors and other birds, Australia might be the place to do it.

Left In the Namib desert the biggest sand-dunes in the world alternate with bare rock and valley bottoms with quite large trees, which provide nesting places for birds of prey.

Above The Lanner Falcon is the most desert-adapted of all big falcons. It preys upon migrants and sandgrouse at waterholes.

Deserts and semideserts

No two ecologists can agree in defining a desert. At one extreme, it is a totally waterless area which receives no rainfall and can support no life whatever. On the other hand, many tracts of what most people would regard as desert receive some rainfall and may even, locally, support quite dense vegetation. Most ecologists would regard the edges of the empty quarter of Arabia, with only about 50 millimetres of very erratic annual rainfall, as a desert. Yet, because much of the rainfall runs off stony uplands and

In naked rock desert south of the Atlas Mountains erosion lays bare the skeleton of the country. Such barren land still supports Bonelli's Eagles, Peregrine Falcons, and Egyptian Vultures.

concentrates in valleys and depressions there are extensive areas with quite a dense growth of grass and shrubs, even a few good-sized trees. Such areas may support ten or twelve small birds per hectare, large lizards, and small mammals, while the carcasses of gazelles, oryx, goats, or camels provide some carrion. Evidently, such a desert could support some birds of prey, while the Kalahari Desert, with 125 millimetres or more of rainfall, covered with low shrubs and grass, and with big, old acacia trees in watercourses, supports quite a number and variety of birds of prey, yet no-one compelled to march 30 kilometres across the Kalahari would call it anything but desert.

Therefore, I will define deserts as areas receiving less than 125 millimetres of annual rainfall and semideserts as those receiving 125 to 250 millimetres. The difference is one of degree and although such areas embrace quite a wide variety of ecological types they are harsh environments. Often, the uplands have long been denuded to rock by occasional and irregular torrential rains and violent winds. The soil may concentrate in depressions forming shifting, wind-blown dunes. Such depressions may retain enough run-off water to grow perennial vegetation, and a desert traveller can often camp in the shade of quite a respectable acacia, doum-palm, or spreading cactus.

Most deserts and semideserts lie between subtropical steppes and wooded hilly country on one side and tropical savanna on the other. They are normally found within a warm, dry climatic belt between the subtropics and the tropics but this is broken by mountain ranges and sometimes affected by cold ocean currents, as in Peru and south-west Africa. Approaching the fringe of the desert, trees and bushes become smaller, more scattered, and reduced in species to a few specialized, aromatic, thorny, or succulent types adapted to the conditions. Perennial grasses and any remaining big trees are now concentrated along drainage lines. Hillsides become naked rock, exposing the geological skeleton of the country. Deserts are fascinating places but there is no doubt that they are harsh and difficult to live in.

They are not nearly so harsh as the Arctic tundra, however, because they are never really cold. Winter frosts are common in the northern Sahara or the Kalahari, but they are not the frosts of Baffin Land, and by day the sun shines. Prey animals, jerboas, kangaroo rats, lizards, snakes, larks, or sandgrouse do not just disappear altogether as in the Arctic. They often take refuge from the intense midday heat by hiding in the shade of bushes or rocks, in holes, or by burrowing. Thus, nocturnal predators may fare better in deserts than diurnal hunters because their prey emerges at night. Some small mammal predators, such as the Fennec Fox, occur only in deserts but no diurnal birds of prey are exclusive to deserts. All also occur either in subtropical steppes or mountainous areas, or in warm, tropical savannas with more than 250 millimetres of rainfall, and have adapted to a harsher climate.

Thirty-three species breed in a desert somewhere in the world. Not one is characteristic of deserts, though a few are undoubtedly taking advantage of the peculiar circumstances of desert to live. Thus, the Sooty Falcon breeds in the midst of the scorching Libyan desert. It is, however, typically an inhabitant of small Red Sea islands and here is treating the habitat as a sort of dry sea in which small, migrant birds are compelled to seek shelter in places where they can be caught. The Lanner Falcon in deserts also preys heavily on migrant songbirds; and both the Lanner and the Peregrine hunt small birds, doves, and sandgrouse arriving to drink at scarce waterholes.

Raptors adapted to desert are of two main classes, those from subtropical woodlands or steppes, and those characteristic of drier tropical savannas which also colonize deserts. There are also a few adaptable, temperate species such as a race of the Red-tailed Hawk breeding in Arizona deserts and the Golden Eagle in the mountains of the northern Sahara. These are, respectively, characteristic of temperate deciduous woodland and of temperate moorlands and mountains. Further north, both escape Arctic winters by migrating but in deserts they are pushed to the limit of their adaptability. The Red-tailed Hawk feeds on snakes and lizards as well as the small mammals it prefers; and the Golden Eagle must subsist at least partly on the large, fat-tailed *Uromastix* lizard. They can adapt their food preferences to survive.

Twelve desert-breeding species are originally subtropical and ten are typically tropical savanna species which colonize deserts. Others occur in tropical savannas and in subtropical steppes or woodlands and four are adaptable subtropical and temperate species. The Peregrine Falcon, which breeds sparsely in deserts, and has evolved a desert race in North Africa, occupies almost any habitat except dense forests. The Red-necked Falcon normally prefers humid savanna with *Borassus* palms but in the Kalahari and in the Transvaal unexpectedly breeds in old Cape Rooks' nests in acacia trees.

Ten desert species are primarily European or Asian, another twelve primarily African, four North American, and four Australian; only one is South American and one cosmopolitan. These figures reflect the largest tracts of desert in the Sahara and the deserts of north Africa, Arabia, and interior Asia. The African desert species originate mainly in tropical savanna rather than the north African desert fringe.

The special interest of the desert lies in investigating how raptors typical of easier habitats are adapting to harsh desert conditions, how they must vary their way of life, exploit new situations, and change their prey preferences. A Golden Eagle studied in the northern Sahara would certainly give a completely misleading idea of the species' behaviour in a typical temperate mountain range, yet is still a Golden Eagle. The wonder is that it can adapt to such extremes as Alaska and the Sahara at all.

Tropical savannas

The change from desert to tropical savanna is clearest moving south from the Sahara into west Africa. The climate remains scorching hot but gradually the country becomes better vegetated. Perennial grasses and trees spread from main watercourses up tributaries and then over slight depressions where water collects. Eventually, most of the ground is covered with grass interspersed with bushes and low trees, and the watercourses are clearly marked by a line of dark-green acacias and doum-palms. This very gradual transition brings us to the fringe of the savanna proper. Similarly, in South America or Australia; there is no abrupt dividing line between semidesert and savanna.

Moving on into tropical savanna, the conditions gradually change from drier to more humid types. The grass grows longer, with more tall perennials. Trees become bigger and the fine, feathery leaved acacias are replaced by other broadleaved species such as (in Africa) *Combretum* and *Terminalia*. In Australia, the acacias have no thorns and the broadleaved trees are still mainly eucalypts; but the general appearance of the countryside is remarkably like savanna in Africa or South America.

Briefly, savanna may be defined as grassland with scattered deciduous trees which never form a complete canopy. If trees are few or absent, savanna becomes grass plains, such as the Serengeti or the seasonally flooded Llanos of the Orinoco. The presence or absence of trees is often related to drainage. Where trees are numerous enough to suppress most of the grass the area becomes dense woodland or thorn bush. Various terms such as 'wooded grassland' or 'orchard bush' are used to describe intermediate types.

The basic characters of savannas, however, are: a ground carpet of perennial and tall annual grasses which dries in the dry season and then readily burns; deciduous trees with corky, fire-resistant bark which drop their leaves in the dry season, in response to drought, not cold; and an intermediate layer of scattered shrubs about a metre high, most of which are burned back in the dry-season grass fires and then spring again from the root stocks. Such vegetation covers at least thirty million square kilometres of South America, Africa, tropical Asia, and Australia. It merges gradually with semideserts as the climate becomes drier, but is generally rather abruptly separated from true tropical forest.

Tropical savannas are richer than any other zone in birds of prey, with 112 characteristic species and another eighteen also typical of some other zones. Individuals of some species may also be common, easy to observe, and quite tame. Carrion-eating vultures and some buteos or buteonines, for example, are often the commonest large birds of prey in savannas. The habitat itself may be little changed by man over vast areas, and in Africa, the largest known communities of big wild animals survive in such relatively unspoiled areas as the Kruger, Kafue, Serengeti, and other National Parks. In some such areas, however, such as Tsavo, a huge population of elephants has destroyed most of the large trees and affected the habitat for birds of prey.

Forty-six genera of birds of prey occur in tropical savannas, more even than in tropical forest, with forty-three genera. Only six genera occur exclusively in tropical savannas, however. Another twenty-four genera are characteristic of tropical savanna but extend to other habitats, especially subtropical steppes and woodlands. Savanna species are clearly much less specialized for a single habitat than those of tropical forests.

The six exclusively savanna genera include four African and two South American but no Australian or Asian genera. The African genera include the exquisite little Swallow-tailed Kite, *Chelictinia*, the Hooded Vulture, *Necrosyrtes*, the magnificent Bateleur, *Terathopius* – the classic emblem of African skies – and the small buteonine Lizard Buzzard, *Kaupifalco*. The Hooded Vulture may occur near human settlements in forest, and the Bateleur may appear almost anywhere but breeds only in savanna. The American genera are the Pearl Kite, *Gampsonyx*, and the Spot-winged Falconet, *Spiziapteryx*; both are small and little known. Most South American savanna species occur also in other habitats, notably grass plains or pampas.

Twenty-three other genera, including in all 122 species, are typical of tropical savanna but occur also in subtropical steppes or woodland. Seventy species of these genera occur in tropical savannas and seventy – not all the same – occur in subtropical woodlands or steppes. Only six of seventy typical savanna species occur also in tropical forest. Typical genera common to savanna and steppe include the elanine kites, *Elanus* and *Ictinia*, large vultures, *Gyps* and *Aegypius*, in Asia and Africa, snake eagles of the genus *Circaetus*, harriers, *Circus* (though apparently no harriers breed in tropical African savannas), chanting goshawks, *Melierax*, and the so-called buzzard-eagles, *Butastur*. Of twenty-seven members of the genera *Buteo* and *Parabuteo* twelve occur in savanna and seventeen in subtropical steppe and woodlands. Both the magnificent Martial Eagle, *Polemaetus*, and the Secretary Bird, *Sagittarius*, are typical savanna birds. Pygmy Falcons, *Poliohierax*, and true falcons, *Falco*, are also commonest in savannas. Of thirty-seven *Falco* species, twenty-three occur in savannas and twenty in the subtropics, again with some in both. Some small savanna falcons have very limited, inexplicable distribution. The Red-necked Falcon, Dickinson's Kestrel, and the Madagascar Banded Kestrel all prefer areas where fan palms (*Borassus* spp) are com-

mon, apparently avoiding similar areas without palms.

Characteristic savanna birds would also include members of some genera preferring other habitats. Several large- and medium-sized eagles of the genera *Aquila*, *Hieraaetus*, and even some *Spizaetus* are quite typical. In south American savannas, caracaras and milvagos augment the scavenging efforts of neotropical vultures. In the Old World, the Black Kite is everywhere, abundant even in Australia. Buzzards and their near allies, such as the Savanna Hawk, *Heterospizias*, and the Bay-winged Hawk, *Parabuteo*, are often the commonest raptors seen. The species occurring are often woodland types which can adapt easily to more open country.

In the three largest genera of birds of prey, *Accipiter*, *Buteo*, and *Falco*, of forty-seven sparrowhawks and goshawks (mainly forest birds) ten prefer tropical savannas and thirty-two prefer tropical forest. Some accipiters can adapt to savanna, and their near relatives the chanting goshawks are mainly savanna birds. Eleven of twenty-six buteos which prefer to hunt in open country occur mainly in tropical savanna but only two are typically birds of tropical forest and these spread into savanna. Of thirty-seven true falcons, twenty-one are characteristic birds of the tropical savannas but only seven breed regularly in tropical forests. All of these catch their avian or insect prey outside forest above the canopy or in adjacent open areas.

The breeding species of tropical savannas, especially in Africa, are greatly augmented by migrants from temperate climates in winter. Generally, these fly right over any

White-backed Vultures are the commonest large vultures of wooded savannas.

intervening, inhospitable desert belts or forests and enjoy a relatively easy winter in a warmer climate than that of their summer breeding haunts. Some such migrants cross the equator. For instance, Swainson's Hawk, the European Steppe Buzzard, the Hobby, Red-footed Falcon, and Lesser Kestrel apparently seek conditions in the summer of the southern hemisphere resembling those of their breeding quarters but they do not breed although they are present quite long enough to do so. Sometimes they enormously outnumber any of their near relatives in their winter quarters. The Steppe Buzzard and Lesser Kestrel, for example, are far commoner in South Africa in the austral spring and summer (October to March) than are the resident Common Kestrel and Jackal Buzzard. Thus, in tropical savannas, the winter migrants from temperate climates have a very important share in the overall effect of raptors on their prey.

Within the tropics, where the climate is always warm or hot, the need to migrate is naturally less than in Arctic or temperate climates. Tropical savannas, however, also experience alternate feast and famine, here caused by alternating rainy and dry seasons. The rainy seasons normally occupy about four to six months from May to September in the northern tropics and November to March in the southern tropics but near the equator there are two three-month dry and wet seasons each year. In the rains, the grass grows long and lush, the trees burst into leaf, small birds breed, and there is a season of burgeoning life far more concentrated and dramatic than the northern spring. In the dry season, the grass dies back and, when it is sere and yellow, someone or lightning sets it alight, and a holocaust of flame passes through the savanna, leaving the ground bare and black, the trees leafless and the shrubs killed down to their rootstocks. The contrast between the black, ash-covered ground of the dry season, baked brick-hard by the scorching sun, and the lush, moist green of the rains is as extreme in its way as is the difference between the naked trees and snow of winter and the deep green leaf of summer in temperate climates.

At least seven species of tropical savanna raptors are regular migrants, and at least another twenty are partly migratory or more or less nomadic. Six of the regular migrants, the African Swallow-tailed Kite, the tropical African race of the Black Kite, the Shikra (which does not migrate throughout its range), the Grasshopper Buzzard-eagle, the African Red-tailed Buzzard, and Wahlberg's Eagle are African. Of these, Wahlberg's Eagle alone crosses the equator; the others migrate to and fro between drier savannas bordering the Sahara and moister savannas near the equatorial forest.

The only South American migrant is the largely aerial Plumbeous Kite, a close relative of the Mississippi Kite of temperate to subtropical steppe and prairie. Its movements are little recorded and probably not all members of the species migrate, though those breeding on the southern fringe of the range apparently migrate north and those breeding on the northern edge migrate south, presumably at different times of year. In Australia there are no regular migrations and more or less irregular nomadism seems typical of the semi-arid, extremely capricious climate.

Comparing tropical savannas with wooded semitropical areas to the north or south, it is again clear that the nearer the equator the less the need to migrate. In tropical savannas only seven of 112 typical species are true migrants (about one in sixteen), whereas in subtropical woodlands twenty out of seventy-three or more than a quarter migrate. The tropical savannas are also a much more important wintering area for northern and subtropical migrants than subtropical areas, partly because many north-temperate species migrate straight through into the tropics to winter, and partly because there are far fewer species in any temperate zone.

To reduce the complexity of the identification problem in tropical savannas, it helps to split up the total into continental areas. Of 112 characteristic species, sixty-one are found in Africa, some of them also in Asia or Europe. Twenty-six occur in Asian savannas and jungles, twenty-one in Australia, and thirty-four are confined to America. Also, in different parts of African savannas, not all of the sixty-one species that might occur would ever be seen. Generally, the maximum number of residents you could expect to see anywhere in Africa is between twenty and thirty with about ten to fifteen winter migrants also. This is difficult enough, perhaps, for anyone used to the comparatively few species seen in north-temperate woods or moorlands, but still it makes the problem of studying the complex habits of savanna species a little less daunting.

In certain continents, certain groups of these savanna inhabitants have proliferated, such as buzzards and their near relatives in South America. Of nineteen savanna species of buteos and buteonines, twelve are found only in America (including the Galápagos islands as American). Five are African, one confined to Madagascar, and only two oriental. There are no buteos in Australia, where certain large kites have evolved to fill the empty buzzard niche. Of four large, scavenging (milvine) kites, three are exclusively Australian while the Black Kite also abounds there. Australia lacks large vultures but there are four each in South America and tropical Asia, and seven in Africa. Four out of five savanna snake eagles are African; the one Asian species is a forest bird that spreads into savannas. All three chanting goshawks are African as are four out of ten savanna accipiters. The three Australian savanna accipiters are not confined to this habitat but the Red Goshawk, *Erythrotriorchis*, mainly inhabiting tropical savannas is a large, aberrant accipiter type more like a small buzzard. In the largest savanna genus, *Falco*, with twenty-three species, nine are purely African (including Madagascar and other islands), three occur in Africa and elsewhere, four are American, and two oriental. The four Australian savanna falcons are again not confined to savanna. In Australia, in the absence of buzzards and vultures, specialized members of other groups, a big accipiter, two large kites, and a large eagle have evolved to fit at least partially these feeding niches.

Any such analysis shows that if there is an unoccupied ecological or food niche, unrelated raptors may evolve to exploit it. New and Old World vultures are unrelated but feed on carrion, while caracaras, related to falcons, are half vulture, half crow. In the savanna, also, the number of species and the diversity of their habits depend more on the total area of the habitat concerned than on the variety of ecological types contained within it. In Australia, with a great variety of grassland, woodland, and bush there are far fewer savanna species than in Africa and none is truly exclusive to tropical Australian savannas.

Tropical forest

The change from tropical savanna to tropical forest is normally abrupt. Out in the savanna the sun blazes down, the ground is probably bare and ash covered, and the gnarled, fire-resistant, low trees are leafless. Entering to seek shade, you push through a dense, protecting screen of shrubs and creepers, and within 45 metres, you are inside a cool, damp, gloomy place where the vertical trunks of huge buttressed trees soar up above an understory of smaller saplings and trees, itself as tall as the biggest savanna trees. The ground is covered not with ash but a layer of leaves and your progress is soon obstructed by leafy shrubs and trailing creepers.

In the rains the contrast is less extreme because the savanna trees are then in full leaf and it is relatively cool and shady outside the forest. You thrust laboriously through waist- or even head-high grass to the forest edge, however, and find that once inside, you can walk in open spaces with no grass at all. The essential contrast is not between trees and no trees, as in the north, but between a mixture of small, gnarly trees and tall grass in bright sunlight to quite different tall trees and no grass in the deep shade of the forest.

In tropical forest, the temperatures vary little; the normal range is from 18 °C to 32 °C. Frost is unknown, and though there are dry and wet seasons, their effects are reduced because true tropical forest will not develop except in deep soil, in at least 1250 millimetres of rainfall, and where at least 25 to 50 millimetres of rain falls every month. It is often said that forests create rain but most forests are the result of rather than the cause of rainfall.

Tropical forests still cover vast areas astride the equator in South America, Africa, and tropical Asia, extending to New Guinea and northern Australia. The largest contiguous areas are perhaps eight million square kilometres in South America but the great forests of the Congo basin are almost as impressive. Equally fine, high rainforest develops in Malaysia and Indonesia but of quite different tree species.

The tropical forests have one fewer characteristic species (111) than tropical savannas and six fewer overall (124) because only another thirteen species occur in this and other habitats. The forest also has fewer genera, forty-three compared with forty-six, but eighteen genera are confined to forest. Of these, fourteen are monotypic with one specialized species each, two have two species, and two five each. These last could not be more different but each is a typical forest genus. The five strange, forest falcons, *Micrastur*, live in the dense forest undergrowth and will not voluntarily emerge into the open at all. The five tiny falconets of the genus *Microhierax* are aerial insect hunters above the forest canopy but breed in old barbet and woodpecker holes and depend on big forest trees for vantage points.

The eighteen genera found only in tropical forests vary from the world's most formidable bird, the huge Harpy Eagle, and its near rival the Philippine Monkey-eating Eagle to tiny falconets, the smallest of all birds of prey.

Several specialized kite genera include the Hook-billed Kite, *Chondrohierax*, which eats arboreal snails, and the Double-toothed Kite, *Harpagus*, related to cuckoo falcons but resembling an accipiter. Three tropical eastern honey buzzards and two forest snake eagles of Africa and Madagascar live only in deep forest; no-one has seen the Madagascar Serpent Eagle for years. The African Long-tailed Hawk, *Urotriorchis*, is a specialized accipiter type with a very long tail. Four very large buteonine eagles (*Morphnus*, *Harpia*, *Harpyosis*, and *Pithecophaga*) inhabit forests in South America, New Guinea, and the Philippines, and two large booted eagles with feathered tarsi (*Ictinaetus* and *Oroaetus*) live in Asian and South American forests respectively. Finally, there are the aberrant, snake-eating Laughing Falcon, *Herpetotheres*, the strange forest falcons of the undergrowth, and the tiny aerial falconets.

These exclusively forest-loving raptors are so interesting that it is strange that they have been so little studied. Of eighteen genera, however, fourteen are very little known, four moderately well known. The only really well-known species preferring but not exclusive to forest is the big African Crowned Eagle, *Stephanoaetus*. Our ignorance is compounded by the difficulties of the environment itself, and the fact that fifteen of the eighteen genera are South American, oriental, and Australasian, areas where active ornithologists are scarce.

Another fourteen genera with ninety-four species prefer forest. These genera also occurring in savanna comprise eighteen typically savanna and seventy forest species. Excluding *Accipiter*, with thirty-two forest and ten savanna species, the relative numbers become thirty-eight in forest and eight in savanna. Either way the forest species outnumber the savanna species two or three times. Several forest genera with a few species ranging into savanna are primarily forest birds. They include the highly coloured King Vulture, *Sarcorhamphus*, cuckoo falcons, *Aviceda*, honey buzzards, *Pernis*, Crested Serpent Eagles, *Spilornis*, and the curious, very little-known, but apparently reptile-eating hawks of the genus *Leucopternis*. Many of these seldom leave forest but others hunt on its edges.

Far more raptors will leave forest to hunt than will enter it from open country, further emphasizing the exclusive nature of the forest habitat. Only about seventeen of 112 savanna species will enter forest; and of these, four, the Black Kite, and three vultures, are found inside forests only around human settlements, seeking scraps. Vultures, hunting by sight, cannot usually locate carrion in forest at all. The rest are birds which, though they live in savanna, prefer forest. Not a single true savanna species usually enters forest. On Reunion Island, however, the resident race of the familiar Marsh Harrier is the only resident raptor and hunts in the forest and in open country, and the Mauritius Kestrel is an extremely rare bird of the forest.

In contrast, at least thirty-six species preferring forest will emerge to hunt and may be commoner in savanna. The common, but little-known, King Vulture, *Sarcorhamphus papa*, prefers forest but also forages outside. All cuckoo falcons (or bazas) are primarily forest birds which like hunting in open spaces or on verges as do several forest snake eagles and harrier hawks. Both buteos liking forest prefer to hunt outside it and several forest hawk-eagles will take advantage of open country. The American Swallow-tailed Kite and the tiny eastern falconets are essentially aerial but dependent on forests, and all five true falcons breeding in forest, excluding the very aberrant and rare Mauritius Kestrel, catch most of their prey outside or above forest. The Mauritius Kestrel feeds much on forest lizards but is now so beleaguered that its original habits may have been obscured.

The most specialized forest dwellers include accipiters and their near allies; of thirty-one forest accipiters, only about four emerge readily. The huge Harpy Eagle and its near relations, and several small eastern hawk-eagles stay inside the dense growth. Finally, the extraordinary forest falcons will not voluntarily leave their gloomy home at all. A captive bird released in the open first hid in a hole, and when that was closed, crouched in a grass tussock. It would go round a small clearing rather than fly across it as even the most skulking forest accipiter probably would. These observations show that forest environment forces its inhabitants to be more specialized than elsewhere. This, in turn, means that a species specialized for forest life cannot readily adapt to life outside. Any such adaptation would most likely happen in areas where fingers of forest and the savanna intermesh on ridges and in valleys, known as forest-savanna mosaic. Here, a still greater variety of birds of prey may occur than in either savanna or forest alone.

Although forest species are apparently more ready to leave forest than savanna species are to enter it, they can only adapt through evolutionary time to the slow, creeping changes resulting from very long-term climatic cycles. At present, however, tropical forest is being destroyed world-wide by man at an alarming rate; once cut it is normally replaced by scrub, or even savanna, often in less than a century.

The magnificent Philippine Monkey-eating Eagle is at least as much endangered by habitat destruction as by shooting and collecting from which it is legally protected. It cannot quickly adapt to life outside the forest for lack of big nesting trees and monkeys to eat. Its African ecological counterpart, the Crowned Eagle, does, however, emerge into savanna or woodland and will eat other animals besides monkeys. It too, however, would not be able to adapt quickly enough to wholesale forest destruction and would disappear. Essentially, forest species can survive only if large enough tracts are conserved for breeding pairs to maintain normal home ranges or territories.

The tropical forest climate is the most equable of any and the need to migrate is reduced by increasing warmth so that you would expect that few forest raptors would migrate. Also, inside the forest, there is no such alternation of drought and lushness which can occur in savanna only a few hundred metres away. No true forest species migrates at all, as far as we know. Some obvious migrants are largely aerial species, such as the American Swallow-tailed Kite, or species that can also live in savanna. Thus, along the Kenya coast in August the Cuckoo Falcon is much commoner than at other times of year; at that season, it must be a visitor from colder regions further south.

So far as we know, tropical forest is also rather unimportant as a wintering area for migrants from temperate or subtropical climates. One of the few species that may winter commonly in African forests is the Honey Buzzard, breeding in dense deciduous and conifer woodlands in Europe but with tropical forest races in Asia. Skulking accipiters might be expected to winter in tropical forest, but northern European and Asian migratory accipiters seldom cross the equator, and are more likely to winter in savanna. American Sharp-shinned and Cooper's Hawks winter south to Costa Rica, perhaps in true forest. The Steppe Buzzard, breeding in forests or woodland in Europe, winters in open country, woodland, or around the edges of forests, but not properly within them; and the Hobby and Lesser Spotted Eagle overfly equatorial forest to reach southern African savannas. Apart from specialized forest species, birds of prey avoid tropical forest.

Many forest raptors are so little known, however, that some of these statements may later prove to be wrong. Nowhere are birds of prey more difficult to observe than in tropical forest. Those which inhabit the upper canopy will often be invisible from the ground because of the intervening understory foliage. On the forest floor in poor light and with restricted movement and visibility it is hard to locate a raptor even if one is heard calling. When working slowly and quietly through the growth, a hitherto unobserved raptor flies silently away, and often cannot certainly be identified.

Advanced radiotelemetric techniques may be the only method of learning much detail about many forest raptors. It may be possible to trap them, fit them with tiny transmitters, and follow them about with a backpack receiver. Even such radio transmitters do not operate so efficiently inside the forest as in the open, however, because the trees interfere with the signals that the transmitters emit.

The tropical forest environment is thus not only among the richest in total numbers of species, and the richest of any in specialized genera and species exclusive to it, but is the least known and studied of all the world's major environments, and the habitat where the greatest advances could be made by concentrated effort and the use of new techniques and sophisticated equipment. It is certainly very difficult but anyone who can overcome the difficulty will probably find abundant interest in the mysterious, rather daunting gloom of the tropical forest.

Tropical montane environments

In several parts of the tropics, high mountains rear to the snowline. The greatest such area is the Andean chain from Venezuela to Argentina but near the equator there are also high mountains in east Africa, New Guinea, and Borneo. These high mountain areas are extremely interesting, full of strange forms of life but support few unique birds of prey because their total area is not large enough. Most species living there are those from other habitats adapted to these rather searching heights. At above 3000 to 3500 metres, true broadleaved forests usually disappear because, although it may still be very wet, almost nightly frost occurs. Specialized forms of heath replace forests and still higher up, at 3600 to 3900 metres, give place to tussock grass moorland or even to expanses of dry dust with a few spiky plants resembling a cold desert. By day the blazing sun flays the skin of the unwary traveller from the north, while an hour after sunset the frost seizes the land in an iron grip. The climate has been aptly described as summer by day and winter by night. Frequent snow or hail help to maintain glaciers on the highest peaks above about 4500 metres.

The high forests of such mountains are often regarded as a montane zone. For our purposes, the botanically poor mountain types are best regarded as a subhabitat of the lowland tropical forest. Sometimes, as in New Guinea, forest is continuous from lowland to highland, and here the large Black-mantled Accipiter and Buerger's Goshawk are confined to the montane forests. Likewise, on the island of Celebes, the tiny Celebes Little Sparrowhawk avoids competition with three other accipiters in montane forests. In Malaya, with no very high mountains, Blyth's Hawk-eagle is principally a mountain forest bird.

The Rufous-breasted Sparrowhawk and the African Mountain Buzzard occur in montane forests in Kenya and Ethiopia but in lowland subtropical woodlands in South Africa. The African Hobby is a mountain forest species in eastern Kenya (because forest there occurs only as islands on mountains) but frequents moist lowland savannas and forest edges in Uganda and western Kenya. Such species cannot properly be regarded as tropical montane species though locally typical of tropical mountains.

Four Andean species are truly characteristic of mountains in the tropics, and three of these occur nowhere else. The most notable is the Andean Condor. Though most of its breeding range lies within the tropics, it may now be commoner in the southern part of its range as far south as Tierra del Fuego. Gurney's Buzzard, *Buteo poecilochrous*, is a very interesting case as it is barely distinguishable in the field from the Red-backed Buzzard, *B. polyosoma*, and systematists argue whether it is a good species. The Red-backed Buzzard occurs along the whole Andean chain from Tierra del Fuego to Bolivia but Gurney's occurs with it in north Chile to Colombia in the same habitat though perhaps breeding at higher altitudes. It is generally larger than the Red-backed but the largest Red-backed Buzzards occur in the Andean range of Gurney's Buzzard. Detailed field studies are needed to elucidate the relationships of these two buteos which must somehow distinguish each other if they are good species.

The Carunculated Caracara and the Mountain Caracara, closely related to Darwin's Caracara, occur only in the tropical Andes, in the north and the centre of the mountain chain respectively. All these members of the genus *Phalcoboenus* perform on Andean moorlands the functions of Hooded Crows and Ravens on the moorlands of Scotland or Wales, eating worms, insects, eggs, small birds and mammals, and dead large animals.

The Red-backed Buzzard and the Grey Eagle-buzzard, *Geranoaetus*, are also chiefly Andean montane species but extend into the foothills and the open plains further south. Similarly, on east African mountains, the characteristic rat-eating buzzard of the high moorlands, breeding up to 3600 metres, is the familiar Augur Buzzard which is typical of savannas at 1800 metres. The South African Jackal Buzzard likewise occurs in any open country from lowland Cape Province to the 3300 metre mountains of Lesotho. These are essentially savanna or steppe species which have adapted to high mountains. Augur Buzzards soar round the highest peaks of Mount Kenya at over 4800 metres but the so-called Mountain Buzzard does not pass the upper forest edge at 3300 metres.

Any other bird of prey found in tropical montane moorland is either a widespread species which can manage to live here among other species or a visitor merely passing over. Almost any large vulture or eagle could overfly 6000 metre high mountains. Verreaux's Eagle, for instance, breeds at 300 metres in northern Kenya and at 4050 metres in the Bale Mountains of Ethiopia, in semidesert and tropical montane moorland respectively. I have seen a Bateleur hunting rats at 3900 metres and migrant harriers from Europe are often seen on high east African moors. The basically Palearctic Lammergeier, a bird of mountains throughout its huge range, penetrates Africa from Ethiopia south to Lesotho via the high east African mountains. Sherpas apparently call Mount Everest *Chomolungma*, meaning 'birds cannot fly over'; but they do! One of the highest altitude records for birds of prey is held by quite ordinary lowland Steppe Eagles that were found dead on the South Col at 7500 metres. The actual record is of a Rüppell's Griffon Vulture killed by a jet at 11 100 metres near Abidjan, Ivory Coast, 10 800 metres above ground.

During the Pleistocene period, one million to 10 000 years ago, glaciers came right down to 3000 metres on Mount Kenya. Montane heath habitat then covered many thousands of square kilometres in east Africa, compared to a few hundred today. Strangely, no bird of prey evolved to take advantage of this extensive habitat, as so many plants did. Though the moorlands of Mount Kenya or the Bale Mountains, abounding in rats, are perfectly ideal habitat for harriers resembling the tussock grass of the southern Andes or New Zealand where the Cinereous Harrier and Marsh Harrier are apparently plentiful, no harriers breed on them. Perhaps the most intriguing fact about the tropical montane moorland habitats is that they ought to be populated by birds of prey specially adapted to them but often are not.

Aquatic habitats

Water abounds in life except when it freezes solid. Accordingly, it is not surprising that, from Arctic seas to tropical marshes, some birds of prey have specialized in hunting the watersides or even the open waters. They include fish-eating species of several not closely related groups, some that feed on water snails, crabs, and even, in one extraordinary species, oil-palm fruit varied by small aquatic creatures. An interesting aspect of aquatic habitats is that these groups are not automatically found wherever there is water, fish, snails, and crabs. Snail eaters and crab eaters are confined largely to the New World, and though some South American caracaras do eat oily fruits, including those of the introduced oil-palm, only the African Vulturine Fish Eagle is largely dependent on this plant.

Aquatic habitats include the shores and estuarine waters of seas and oceans, inland lakes and rivers, and marshes and swamps. In marshes and swamps, land and aquatic habitats mingle, and marshland birds of prey (mainly harriers) also often hunt over open, dry country; such typically lowland savanna species as the South American Savanna Hawk often likewise feed in marshy places. The shores of big lakes and rivers or of the sea, however, form distinct boundaries between land and water. Here, most birds of prey either exploit the watery environment or dry land habitats nearby. Aquatic habitat may occur in the midst of a semidesert; Pallas' Sea Eagle is apparently as much at home on the lakes of the high plateau of Tibet, as on the Jamuna River or the Caspian Sea.

Most aquatic species feed upon fish. These include the Osprey, the eight sea and fish eagles of the genus *Haliaeetus*, two fishing eagles of the genus *Ichthyophaga*, and the Fishing Buzzard, *Busarellus nigricollis*. Though all eat fish as their staple diet, the sea eagles vary this diet with seabirds and waterbirds, mammals, and carrion, and the Fishing Buzzard also eats lizards, snails, and rodents.

The almost cosmopolitan Osprey is the most exclusive and anatomically specialized fish eater. It does eat occasional birds or mammals but 99 per cent of its food is live fish caught in the open water, rarely picked up dead. Found on seacoasts and inland waters the world over, either as a breeding bird or as a migrant, all it apparently needs is a place to fish and a tree or rock to nest on because it will even breed on the ground. Very oddly it does not regularly breed in Africa south of the equator or in South America. It breeds round the coasts of Australia, however, so that southern latitudes are no obstacle. The reason for this distribution is extremely obscure and certainly cannot be due to food supply.

The sea and fish eagles and fishing eagles are more widely distributed as breeding birds than is the Osprey, occurring from Arctic latitudes on seacoasts to the equator and South Africa, but absent from South America. What prevents sea or fish eagles breeding in South America? The rivers there are bigger than anywhere else in the tropics and certainly abound in fish. Yet there are fish or fishing eagles in tropical Africa, Asia, and Australia in apparently less favourable habitat. The Lesser Fishing Eagle even frequents quite small streams in forests. This is another apparently quite inexplicable quirk of distribution related to the aquatic habitat.

The range of the most aberrant member of the fish eagle group, the African Vulturine Fish Eagle, is practically coincident with that of the Oil-Palm, *Elaeis guineensis*. It occurs in mangrove swamps, along rivers and streams and does eat small fishes, crabs, and even molluscs in such areas. A large river such as the Zambezi in Zambia, however, lacking Oil-Palms, will not do. The Vulturine Fish Eagle apparently must have Oil-Palm fruits and where they abound breeds in forests or savannas far from any large expanse of water. Perhaps it originated as an aquatic raptor but took to eating vegetarian foods at a later time.

In South America, the very variable buteonine group has evolved a specialized fish eater, the Fishing Buzzard, and three large buteonine hawks feeding almost exclusively on crabs. The Fishing Buzzard partially fills the fish-eating niche, not filled in South America by any typical sea eagle or by breeding Ospreys, though both breed in Florida. The much smaller, weaker Fishing Buzzard could scarcely exclude much more powerful, true fish eagles and it must have evolved to eat fish on the spot without their competition.

Crabs abound on seacoasts and coastal marshes any-where. Small fiddler crabs are as abundant in mangroves in tropical America and Africa or Asia, and ghost crabs run along every tropical beach. All crabs present the same sort of problem having hard shells and pincers and displaying a degree of quixotic, indomitable courage in the face of much bigger adversaries which may be mainly bluff but usually works. Yet only in South America have three crab-eating hawks evolved, of varying size and habitat. The Vulturine Fish Eagle, where it occurs away from the range of Oil-Palms may be dependent on freshwater crabs, but elsewhere it only eats a few crabs as a sideline.

Similar geographical specialization occurs in the two South and Central American snail-eating kites, *Rostrha-mus*, anatomically specialized to eat snails. Both eat the same kind of *Pomacea* snails, large, round, and doubtless nutritious of a type perhaps commonest in South America. There are plenty of molluscs elsewhere, however, but no raptor regularly eats them. Again, the Vulturine Fish Eagle eats a few small molluscs as a sideline.

I stick out my neck here to include Eleonora's and the Sooty Falcon as mainly aquatic, because both feed mainly on small birds migrating from Europe and Asia across the Mediterranean and Red Sea on a broad front. Eleonora's Falcon breeds on islands in the Mediterranean, and the Sooty Falcon in the Red Sea and Persian Gulf regions. Only one colony – in Morocco – of Eleonora's Falcon has ever been found on the mainland, but the Sooty Falcon has been found breeding in the depths of the Libyan desert and this may be commoner than is supposed. Both these falcons migrate eastwards round the horn of Africa, pro-bably mainly near the coast, and winter in Madagascar. Neither is aquatic in that it catches its prey in the sea but both are perhaps more exclusively dependent on the sea than even the Osprey because they breed only or mainly on sea islands and their food supply is unfortunate migrants caught over the sea as they make for the very islands which harbour the colonies of their enemies.

A number of other species such as Marsh Harriers are

marginally aquatic, living in marshes, and catching much of their prey in or near water. The Brahminy Kite and Australian Whistling Eagle are links between kites and sea eagles. Even such properly dry-land birds as the Asian Grey Frog Hawk, *Accipiter soloensis*, feeding on frogs, the South American Savanna Hawk feeding mainly on snakes, lizards, and amphibians, and Red-shouldered Hawk of North American woods, which eats many frogs, are partly aquatic. Watery places are not their only or even their favourite habitat, however, but perhaps they are slowly evolving towards aquatic specialization. The Marsh Harrier, especially, is a highly variable species with many races, some insular and capable of adapting to almost any habitat from tall reeds to wheatfields, even forest (on Reunion). In competition with other harriers, as in Europe, it tends to stick to marshes but the same birds hunt quite freely over dry grasslands or even cultivation on migration in Africa.

The aquatic habitat, at least on the seacoast, generally provides food year round, so that no species of bird of prey dependent on it for fish or seabirds need be wholly migratory. North American and European Ospreys do migrate, right down into the southern hemisphere, where no Ospreys breed, but some Osprey populations are resident. The European Sea Eagle, Steller's Sea Eagle, the Bald Eagle, and Pallas' Sea Eagle are also partially migratory though often adults are resident while the young migrate. The Florida race of the American Bald Eagle breeds in the cooler season of the year and migrates northwards perhaps to avoid the hot summer, whereas its Alaskan cousins must migrate southwards to avoid the cold northern winter. Both races occur in the central United States and Canada at different times of the year. Again, not all individuals migrate. Eleonora's and the Sooty Falcon are completely migratory but they are somewhat special cases and perhaps should not be regarded as aquatic birds at all.

To me the intriguing point about the aquatic environment is that birds of prey have responded very unequally to the opportunities it presents for easy feeding and breeding in inaccessible offshore islands. Why do Ospreys breed nowhere in the southern hemisphere but Australia? And why do the otherwise successful and less specialized sea and fish eagles not occur on Amazonian rivers? These are conundrums which we may never be able to answer but they are nonetheless intriguing.

Towns

Large towns are a phenomenon of comparatively recent historical times but moderately large towns and villages have existed for hundreds of years, notably in Europe and

Asia. No large towns existed in either northern or southern America. Nevertheless, some tropical American species have colonized large towns almost as successfully as Asian Black Kites which, because of their willingness to occupy human habitats, are probably the world's commonest and most successful large birds of prey.

Although no bird of prey is exclusively dependent on towns, several are undoubtedly much commoner in towns than anywhere else, and presumably prefer town life. These include the Black Kite, several Old World vultures, especially the Egyptian Vulture, and the New World Black Vulture. In Ethiopia there are more Lammergeiers around towns and villages than in open uninhabited mountains but they do not breed in towns.

The highest known densities of any breeding birds of prey have been recorded in New Delhi, India. In 37 square kilometres, 715 breeding territories were found with 679 occupied nests, an average of 19·3 pairs per square kilometre. Black Kites made up 83 per cent of the total, with 16·1 pairs per square kilometre (about one pair per 6·5 hectares). The next most numerous species were large vultures, mainly White-backed Vultures, with 400 pairs nesting in colonies. There were 100 pairs of Egyptian Vultures (more than in all southern France) breeding on old ruins and temples. Having seen Indian cities where Black Kites literally swarm, these figures do not surprise me. In New Delhi there were about 2400 pairs and at similar densities the urban Indian population of Black Kites must easily exceed a million. At comparable density, London would support 15000 to 20000 pairs of Black Kites.

Similar very high densities are known in African and South American cities. In Kampala, Hooded Vultures average more than four per square kilometre, with about 200 to 210 present, mainly feeding on rubbish dumps, in about 44 square kilometres. In Addis Ababa, I should expect to see more vultures and kites in a day than in a week in the Nairobi National Park, and roadside counts through Ethiopia indicate that scavenging birds round towns and villages form 86 per cent by number of all birds of prey seen. In South America, counts in a variety of habitats in Brazil and Paraguay showed that the Black Vulture, *Coragyps atratus*, greatly outnumbered all other species mainly in towns at very high local densities. Town scavengers combined, including caracaras and milvagos, outnumbered all other birds of prey about eight to one. In any country with a large number of species of birds of prey, a few species may colonize towns so successfully that they outnumber all other species found.

These unnaturally high densities are due to man and his towns. They occur mainly in the east and Africa, but the

South American counts show that a similar situation can develop quite rapidly in hitherto underpopulated areas. Some species are still adapting because in India and north Africa the Tawny Eagle is common in towns, but not in South Africa. No-one should object to scavenging raptors in towns, particularly in the tropics, because here as elsewhere, they perform a very useful function eating all manner of bits of meat and bones which could otherwise be a health risk.

Sometimes, the population of scavenging raptors based on towns is much greater than the actual breeding population. Black Kites, for instance, migrate from Europe to tropical Africa, where the local resident race is also migratory. In Kampala and Nairobi migrant Kites far outnumber the breeding population and forage out from Nairobi at least 20 to 30 kilometres searching the radiating roads for animals killed by cars at night, returning in the evening to roost in certain places. The Red Kite, however, which formerly scavenged in the streets of London and bred at Gray's Inn, has now disappeared as a town scavenger in Europe, though it still scavenges in the Cape Verde Islands. When human sanitary habits improve, the scavenging raptors disappear but in Europe and North America this niche has at least partly been taken over by the swarms of gulls that frequent rubbish dumps.

Most of the town dwellers are scavengers but a few others colonize towns because they provide artificial, cliff-like nesting sites and an available food supply. Several species and races of kestrels breed in towns where they find plenty of rats and mice on waste ground. The Lesser Kestrel nests in several cathedral towers in Spain. The Peregrine Falcon, the Indian Lagger Falcon, and African Lanner Falcon also sometimes breed on buildings and man-made structures. Even occasional Bonelli's Eagles have been recorded breeding on buildings. For these, towns merely provide an artificial, cliff-like breeding site while a relative abundance of domestic pigeons, rats, mice, and smaller birds provides a food supply. They are not true town dwellers like the scavengers but are adaptable and successful, cliff-dwelling birds that have colonized a new, artificial cliff. Their breeding performance, as demonstrated by the celebrated Peregrine that bred for twelve years on the Sun Life building in Montreal, can be greatly improved by the provision of suitable nesting trays or boxes because the unyielding concrete of most modern buildings is a poor substrate for nesting; the eggs roll off.

One specialist feeder is a marginal case. The Bat Hawk, *Machaerhamphus*, is an aberrant, long-winged, swift-flying, crepuscular kite which feeds almost exclusively on bats, mostly caught in a few minutes around dusk as they swarm out of their daytime roosts to feed. Bats typically inhabit caves and Bat Hawks hunt them, together with swiftlets, round the great Niah caves in Borneo, for example. Bats, too, have recognized that human dwellings provide a fair cave substitute, however, and many tropical bats roost abundantly in the roofs of houses. They swarm out at sunset and can be caught by Bat Hawks over any open space such as a playing field, a railway station platform, or a beach. I have seen Bat Hawks hunting over the tourist hotels at Malindi, illuminated by the lights, and on one occasion, at the Nairobi Agricultural show, one was hunting bats, attracted by insects, attracted by the lights of the arena ablaze for a military tattoo. A Bat Hawk is as likely to be seen over the school playing fields in a large town as anywhere.

Towns are, alas, likely to increase in size and extent and quite possibly other birds of prey may, in time, colonize them as an acceptable or preferred habitat. They often support high densities of certain small birds as a result of the varied habitat created by gardens and irrigation in otherwise dry localities. Thus, a new feeding opportunity is created and some other birds of prey should take advantage of this in time. I do not share a prevalent fear, however, that as human population increases the whole world will be covered with vast towns because townspeople are wholly dependent for food on often less numerous farmers. There is much more likely to be a mass die-off of city populations, with a consequent temporary abundance of food for vultures.

Space has dictated oversimplification in this survey of the world's major and specialized bird of prey habitats. In the richer habitats especially, anatomical and ecological specialization means that not all the species listed can conceivably occur, even on one continent. Clearly, however, the warmer the habitat generally, the greater the number of species that may occur, from only four in the tundra to over 100 in each of the main tropical habitats. It follows from this that the best areas in which to study birds of prey are the tropics and that these rich tropical habitats are also those in which the role of ecological or anatomical specialization is best studied.

Unfortunately, however, it has become more difficult to study birds of prey in the tropics in recent years partly because of uncontrollable political trends. At the same time, new areas of the tropics are being opened up and modern transport and equipment can make what would, fifty years ago, have been a daunting if not positively dangerous undertaking, quite easy and comfortable. The greatest opportunities for furthering our knowledge of birds of prey lie in the tropics of South America, Asia, Australasia, and Africa, in that order of importance because tropical African species are better known than those of other tropical areas.

Anatomy, structure, and way of life

Birds of prey share the essential features of other flying birds. They are bilaterally symmetrical, warm-blooded vertebrates, covered with feathers, and have a four-chambered heart. The feathers insulate the warm body from changes in temperature, perform the functions of adornment, perhaps camouflage and even mimicry, and in the wing and tail are highly specialized for flight.

Compared to their nearest relations, gamebirds and ducks, birds of prey are, so to speak, all feathers. A plucked bird of prey would be unrewarding to any cook (though the Honey Buzzard and Eleonora's Falcon are eaten in Mediterranean countries). A cockerel may weigh as much as a Bonelli's Eagle but to handle the Eagle is a wholly different proposition. A male Mute Swan is said to be able to break a human arm and can look formidable but again is a totally different proposition to a Harpy Eagle of comparable weight. Weight for weight, birds of prey are extremely powerful compared to others, and the destructive capacity packed into quite a small, light body is astonishing.

Here I propose to concentrate on how the basic anatomical features of birds of prey are modified by their way of life. The most important modifications are to the head and beak for seeing, hearing, and eating, to the wings for flying, and to the tail for steering. The feet are modified for killing, and are the most dangerous organs when a bird of prey is handled. These various parts can be adapted to the habitat in which the raptor lives; forest, especially, imposes certain forms of wing and tail. Wings and tails especially are also modified by hunting methods, and prey preferences affect especially beaks, feet, and talons. Habitat, mode of hunting, and preferred prey all combine to produce an organism perfectly fitted for its role in nature.

Anatomy should be discussed in a logical sequence, and I shall begin at the head, work from there to the body, and pay special attention to the movable appendages to the body, the wings, tail, legs, and feet. As a type, I shall use the genus *Buteo* (or buteos), generalized, common, and successful birds of prey, adapted to most habitats except dense forest from Arctic to equator and sea-level to 6000 metres, killing a very large variety of prey from beetles to quite large mammals and birds, and taking some carrion. They are not extremely specialized, having moderately

long, rather rounded wings, moderately long tails, fairly strong legs with moderately strong feet and talons. Buteos vary in mass from the Roadside Hawk of about 300 grams to the Asian Upland Buzzard weighing up to 2000 grams but they are all much the same general shape. Most ornithologists can recognize a buteo from an accipiter or a harrier at once and they will serve much better as a type than a more specialized raptor. See Fig. 4.

The European Common Buzzard is near the average in size in this group. Some standard dimensions (in millimetres) for the nominate European race of the Common Buzzard are: **males**, *mass* 535 to 985 grams, averaging 806 grams; *wing* 350 to 400 (375); *tail* 190 to 210 (200), about

Fig. 4 Common Buzzard *(Buteo buteo buteo)* in flight showing general layout and structure of feathers.

A Common Buzzard; it is moderately sized, without exaggeratedly long wings forming 'shoulders'; it has short legs with moderately powerful feet, a moderate-sized beak, and a longish tail.

71

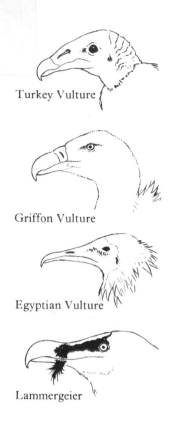

Turkey Vulture

Griffon Vulture

Egyptian Vulture

Lammergeier

Fig. 5 Heads: Turkey Vulture –
head and upper neck bare;
Griffon Vulture – whole long neck
bare, for feeding deep inside
carcasses; Egyptian Vulture –
partially feathered head, not
inserted deep into carcasses;
Lammergeier – feathered head,
bristly 'beard' whose function is
unknown.

53 per cent of standard wing length; *wingspan*, about 1150, 3·1 times the wing length; *tarsus* 70 to 80; **females**, as in most birds of prey, are bigger, *mass* 700 to 1200 grams (938); *wing* 375 to 425 (400); *tail* 210 to 233 (222) about 55 per cent of wing length; *wingspan* about 1285, 3·2 times wing length; *tarsus* 70 to 80. On average the tarsus of the male is shorter than the female's but there is an overlap. Many other buteos have similar proportions, and most weigh 600 to 1200 grams.

A Common Buzzard's head is covered with feathers except for the lores, a space in front of the eye, covered with bristles. The hooked beak is neither very long nor short, and has no serrations or teeth on the upper mandible. The base of the upper mandible is expanded into the fleshy cere, in which are placed the unspecialized, round nostrils. In the Buzzard and many other birds of prey, the cere is yellow in adults though in many young raptors it is duller, grey or green. The precise function of the cere, possessed by only a few other groups of birds such as parrots and pigeons, is not clear but it is sensitive to touch and bleeds readily.

The gape of a Buzzard's beak extends so that a facing bird has the appearance of a grin. This wide gape enables the Buzzard to swallow quite large animals whole. The eyes, situated on the sides of the head just above the base of the gape, are large, round, almost fixed in their sockets, and look both forward and sideways. In the Buzzard and most other birds of prey, they are protected by a bony ridge projecting from the skull above them. This, with the nictitating membrane, or third eyelid which can be drawn at will across the eye, helps to protect the delicate and vital eyes from damage when crashing into vegetation, dealing with struggling prey, or when feeding young.

This general head arrangement is varied in several ways. In both Old and New World vultures, the whole or part of the head and neck is bare of feathers or has only a fuzz of down or short bristles. The primary reason for the bare skin is the carrion-eating habits of vultures which often thrust their heads and necks into small, gory openings. One has only to wash one's hair after taking active part in dissecting a hippo to appreciate the advantage of baldness. Some vultures, such as the Old World griffons which thrust their whole heads and necks far inside carcasses, have the complete head and neck bare. The large New World condors also have most of the head and neck bare, but smaller Turkey and Black Vultures have the neck feathered, and the Old World Egyptian Vulture has only the face and top of the head bare. The Lammergeier, a vulture which specializes in dry bones, has the head fully feathered but sports an extraordinary 'beard' of stiff, downward-pointing bristles on either side of the beak.

Head of a female Goshawk; the piercing yellow eye is protected from damage by the eyebrow or supra-orbital ridge; the beak lacks conspicuous teeth.

Crowned Eagle

Snake Eagle

Honey Buzzard

Common Caracara

This specialized organ has not been explained but perhaps it has a tactile function, preventing the Lammergeier from pushing its rather long beak too far into the hollows of large bones in search of marrow, but that is only my guess.

The bare skin is primarily an adaptation to feeding on carrion but it is modified into bright colours and is often used in display. The colourful American King Vulture, has a bright orange neck, wrinkled yellowish skin behind the eye, a purplish wattle hanging down in front of the eye, and a huge, bulbous, wattled, bright orange cere. My artist friends tell me it is the most 'paintable of all raptors'. Several other New and Old World vultures have bright coloured heads; for instance, the red heads of the American Turkey Vulture and the Asian Black Vulture. The bright yellow face of the Egyptian Vulture is set off by a ruff of erectile, pointed feathers on the head and neck, and in the Lappet-faced Vulture, the bare skin is folded to form lappets like ears. In others, the bare skin is dull but colours may change in emotion. The pink-faced Hooded Vulture flushes brighter red in emotion or threat and in a displaying Andean Condor the dull reddish skin of the head and neck turns bright yellow as the male bows and arches his head and neck towards the female. Thus, a primary function, derived from the need to feed on gory carcasses, has been secondarily modified in various colourful ways for

Fig. 5 cont. Crowned Eagle – powerful beak, erectile crest used in display; Snake Eagle – large, brilliant yellow eye, cowled head; Honey Buzzard – lores tightly feathered, perhaps to protect against wasp stings; Common Caracara – partly bare face associated with carrion feeding.

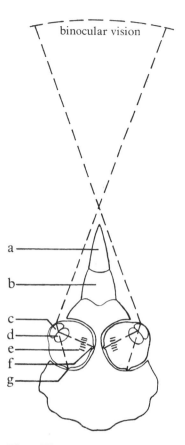

binocular vision

a
b
c
d
e
f
g

Fig. 6 The structure of a bird of prey's head and eyes, showing binocular forward vision. Key: (a) upper mandible; (b) cere; (c) cornea; (d) lens; (e) pecten; (f) lateral fovea; (g) temporal fovea.

display, either in threat or in marital approaches.

A few other carrion-eating birds, such as caracaras, have partially bare faces as do the African Vulturine Fish Eagle (hence the name) and the African Harrier Hawk. In the Vulturine Fish Eagle, the bare face may be associated with eating the messy fruit of Oil-Palms which Harrier Hawks also sometimes eat. In the Harrier Hawk, however, the bare yellow or greenish face blushes pink or red in threat, or when a male and female are on the nest together. In such cases the primary function of a bare face may now be less important than its function in display.

Of two types which eat venomous wasps, one, the American Caracara, has a bare face, the other, the European Honey Buzzard (and its relatives in tropical Asia) has the face covered with tight, scale-like feathers. The closely feathered lores of the Honey Buzzard supposedly protect it from the stings of angry wasps. Frenzied wasps attack and sting other animals including man when their homes are being despoiled by the Honey Buzzard but have little deterrent effect on the bird itself. The American Caracara has no such protective shield of hard, stiff, little feathers but it possibly exudes some kind of chemical repellent which deters the angry wasps from stinging its bare face.

The feathers of the crown of the head are often elongated as a crest, normally erected in threat display, so that a large crest undoubtedly enhances such displays. Crests are also, however, erected by many species in nuptial display or excitement at the nest and are not specifically connected with threat.

While some very large, aggressive species, such as the Harpy Eagle and the African Crowned Eagle have crests, so do some quite inoffensive, insectivorous birds such as cuckoo falcons and some Honey Buzzards while the most impressive crest of all is the long, floppy plume of the African Long-crested Eagle, a rather inoffensive rodent eater. Because some races of the Changeable Hawk-eagle are crested and others are not, it is difficult to say what function crests may serve and they are clearly not vital for survival. The generalized, adaptable, and highly successful buzzards have no crests and nor do falcons, though a few accipiters do.

The eyes of diurnal birds of prey are absolutely vital to their existence and often strikingly beautiful. A Common Buzzard's eye is almost the same size as mine, though I am about 58 times as heavy. The large, round eyes are nearly immovable in their sockets though small lateral adjustments may be possible. To see perfectly in any direction, a bird of prey must turn its head; it then has binocular forward vision like ourselves. To see behind it, it can swivel the head on the very flexible neck.

A bird of prey has binocular vision through about 35 to

50 degrees of arc and lateral vision with one eye on either side through about 150 degrees varying somewhat in different species. The blind spot, where they cannot see without turning the head, is about 20 degrees. Ospreys lacking a supra-orbital ridge seem to have unusually prominent, bulging eyes which perhaps denote a greater rearward field of view. No bird of prey can look 'under its nose' like a Bittern, behind it like a Woodcock, nor can they converge their eyes to look forwards or backwards as can some cuckoos.

Within the field, birds of prey have very acute vision, partly because of the size of the image cast on the retina of the very large eye, and partly because of the density of the sensitive visual cells or cones. The resolving power of the retina is about three to four times that of man. Moreover, the eyes of birds of prey have two pits, or foveae, one directed sideways with monocular vision and one directed forwards with binocular vision. In these the cones are still more concentrated. The Buzzard's foveae have a million cones per square millimetre suggesting visual acuity about eight times that of man. I can believe this because I have seen an Augur Buzzard catch a small, green grasshopper among green weeds at 100 metres. A similar grasshopper, placed in plain view on top of a post, fades out of my sight altogether at 30 to 32 metres.

These basic details are modified in certain species. Bat Hawks, hunting bats in the dusk, have very large eyes, as does the Congo Serpent Eagle living in the gloom of the equatorial forest. Forest falcons, living in even gloomier undergrowth of tropical forest and moving about freely at dusk, have large but not enormous eyes; perhaps this is because they hunt partly by ear.

Though no diurnal raptor has the magnificent, glowing orange orbs of an Eagle Owl, the eyes are often very striking, bright yellow, orange, even red. Those of the Common Buzzard, most other buzzards, and many other birds of prey, are light brown. Immatures often have brown or grey eyes when the adults' eyes are yellow. Young Hen Harriers can be sexed in the nest on eye colour alone, the males having brighter eyes than females, which they retain through life. A bright yellow or red eye is often, but not always, a sign of maturity; young snake eagles have bright yellow eyes when they leave the nest.

The groups with the most striking yellow or orange eyes are snake eagles, fierce, bird-killing accipiters, and some swift, small eagles such as Ayres' Eagle, which lives on birds. Several large, powerful eagles such as the Golden or Verreaux's Eagle have brown eyes but those of the bird-eating Martial Eagle are yellow, and those of the Philippine Monkey-eater almost blue. Most non-predatory vultures have brown eyes but the Lammergeier and the American

Andean Condor

Snail Kite

Common Buzzard

Sparrowhawk

White-tailed Eagle

Fig. 7 Beaks: Andean Condor –
huge wattle on cere, powerful
beak for tearing large bodies;
Snail Kite – highly specialized
long, curved 'tool' for extracting
snails; Common Buzzard –
moderate-sized, unspecialized
beak; Sparrowhawk – short beak,
suited to bird eating; White-
tailed Eagle – very large,
powerful, yellow beak.

King Vulture have white eyes surrounded by red rims. The male Andean Condor has a brownish eye and a female's is red, like garnet. Most swift, bird-killing falcons have full, dark eyes often surrounded by yellow eyelids. The Greater Kestrel, a relatively sluggish species, has a white eye.

It is tempting to associate yellow or red eyes with ferocity or keen sight. Most snake eagles, needing very keen sight to see their camouflaged prey in cover, have yellow eyes, although the Bateleur has mild brown eyes but does not depend on snakes. Some bird killers have yellow, others brown eyes. Some insectivorous species such as the Honey Buzzard or cuckoo falcons have prominent, bright yellow eyes – but they too may need keen sight. The Black-shouldered Kite hunting grass mice in open plains and some forest accipiters have red eyes. The yellow eyes of some accipiters and Ayres' Eagle darken with age to orange or even red, so that this also must be related to recognition. Presumably brown or grey-eyed immatures need keen sight just as much as their yellow-eyed parents and again this must be related to recognition. Any careful analysis results in so many apparent contradictions that while we may suspect that eye colour may affect keenness of sight, it also clearly has other functions, such as recognition. New facts about the number of visual cells in the retinas of yellow- and brown-eyed raptors could help to solve such problems.

The beak can vary from the blunt, almost chicken-like beak of some caracaras to the enormously powerful, arched beaks of some sea eagles and vultures, and the exaggerated curved beaks of Hook-billed and Snail Kites. Shape and strength of the beak in most cases are related to the mode of feeding and the type of prey consumed. In the caracaras of the genera *Daptrius* and *Phalcoboenus*, the chicken-like beak is almost unhooked. The more predatory *Polyborus* caracaras and milvagos have stronger, more hooked beaks, and perhaps eat more carrion and fewer insects and grubs though little detail is known on this subject.

The beaks of all other birds of prey have a hooked upper mandible lying over the lower mandible. The unspecialized beak of the Common Buzzard can be taken as a norm from which others diverge. Beaks of the bird-eating falcons and accipiters are relatively shorter, and those of the large, powerful, mammal-eating eagles are relatively heavier and longer. In most species, females have larger, more strongly arched beaks than males, perhaps because they can kill and feed on bigger prey.

The Philippine Monkey-eating Eagle and Steller's Sea Eagle, both enormous, have exaggerated, laterally flattened, almost grotesquely arched upper mandibles. In the Sea Eagle the bill is also bright orange-yellow, like the large cere. The adult European Sea Eagle also has a yellow bill but it is less arched. Possibly these exaggerated, arched bills may again be adaptations for display rather than feeding. The lateral flattening is supposed to help the bird see past its own rather grotesque beak when using binocular vision. It is unlikely that these beaks are needed for feeding because the even larger Harpy Eagle and the African Crowned Eagle, both feeding on large mammals, lack such exaggerated beaks.

Hook-billed and Snail Kites have the upper mandible greatly elongated, sharply hooked, and pointed, adapted for feeding on snails. The Hook-billed Kite feeds on arboreal and land snails including the genera *Polymita* and *Strophoceilus* but how it extracts the mollusc is still unknown. The races of the Hook-billed Kite vary in the size and curve of the beak, perhaps related to the various molluscs they feed on. The beak is perhaps not used in quite the same way as that of the aquatic Snail Kite, which feeds exclusively on water snails of the genus *Pomacea*. In this case, we now know that the sharp, curved point of the beak is thrust round the helix of the snail's shell to cut the columellar muscle, so freeing the mollusc to be eaten. Several other mollusc eaters, such as the Open-billed Stork and Limpkin, also cut the columellar muscle to extract the food.

If the field study of beaks in relation to preferred food can reveal reasons for specialization, then the shape of the beak should suggest food preference where these are unknown. The beaks of six species of vultures occurring together in the Serengeti Plains are clearly adapted to their different functions, and the same principle should be applied to the less known New World vultures. Attempts have been made to deduce the food of little-known accipiters from the shape and relative strength of the beak, sometimes combined with that of the feet. Such deductions are not infallible, however, unless supported by good details of the actual food taken.

All falcons, and their relatives the caracaras, the Laughing Falcon, and forest falcons, have two serrations or 'teeth' on the upper mandible. True falcons use these teeth to break the necks of their bird or mammal prey after catching them in their feet; they are the only diurnal raptors that do this. Such 'toral teeth' as they are called, are, however, also possessed by several inoffensive kites. Cuckoo falcons, bazas, and the Double-toothed Kite (*Harpagus*) have two. The Double-toothed Kite resembles fierce accipiters, but apparently feeds largely on insects and lizards. Cuckoo falcons (misleadingly named because of their apparent resemblance to cuckoos in plumage and to falcons in the teeth on the bill) are mainly insectivorous but eat some lizards. Most insectivorous and lizard-eating birds of prey

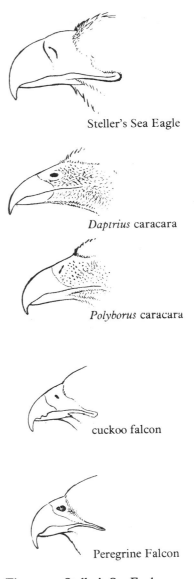

Steller's Sea Eagle

Daptrius caracara

Polyborus caracara

cuckoo falcon

Peregrine Falcon

Fig. 7 *cont*. Steller's Sea Eagle – exaggerated, very deep beak, purpose unknown; *Daptrius* caracara – scarcely hooked beak, omnivorous; *Polyborus* caracara – more strongly hooked, predatory beak; cuckoo falcon – beak with two marked teeth on upper mandible, purpose unknown; Peregrine Falcon – marked tooth on upper mandible used for breaking neck of prey.

Fig. 8 Beak of Turkey Vulture showing pervious nostril typical of New World vultures.

such as the Secretary Bird and the chanting goshawks lack any such teeth, so that I do not see why cuckoo falcons and the Double-toothed Kite should need them. Cuckoo falcons do, however, use their beak to cut off green twigs, as neatly as a pair of secateurs, when making their nests but it seems hardly credible that such an adaptation could evolve for that reason alone.

The nostrils are situated in the cere and they are adapted in various ways. In general, birds of prey have no sense of smell and the nostrils are for breathing alone. The New World vultures have pervious nostrils, however, and the olfactory organs are much larger than in other raptors. Even they apparently have rather a poor sense of smell, because in experiments, they cannot always find stinking carrion near them. The Turkey Vulture does, however, have some sense of smell, and can locate leaks in gas pipe lines by means of a small proportion of a vile-smelling compound added to the gas. The King Vulture, an exceptionally interesting but very little-known species, is the *only* vulture which apparently prefers forest. You might deduce that it could be able to detect putrid prey within forest by smell alone. In South America there are few powerful mammalian scavengers such as hyenas which, in tropical Africa, would get there first.

The slit-like nostril of the Osprey can be closed at will, clearly an adaptation to the Osprey's method of catching fish by a terrific vertical plunge and splash which may partly or wholly submerge the bird. The Osprey plunges feet first and does not dive like a Gannet so that this adaptation clearly helps to prevent getting water up the nose. If you jump off a high diving board feet first, you hold your nose!

Some birds of prey, notably the swift, bird-killing Peregrine Falcon, have bony tubercles inside their nostrils. These are supposed to act as baffles in high-speed flight to prevent rushing air from entering the lungs. Evidently, a stooping Falcon must continue to breathe normally so that such an adaptation would be an advantage either in hunting or the equally spectacular, often prolonged, aerial evolutions of nuptial display. If, as some claim, however, a Peregrine can only dive at 38 metres per second, it evidently does not need any such special adaptation (for even I, with my big nostrils and flabby human face, can cope with a 44 metre per second gale).

The ears in diurnal birds of prey are not obvious or specially adapted to pinpoint sounds, as owls' can. The ear in a bare-headed vulture is just a little round hole, but in most birds of prey it is covered with, and presumably muffled by, feathers. Birds of prey, however, do have acute hearing, though their frequency range may be an octave or so less than ours.

In dense forest, hearing becomes important to locate prey, and in the forest falcons the feathers of the ear region are small, stiff, and upcurled creating a slight ruff. Their ear openings are large and they clearly hunt partly by ear; they can be attracted to within 5 metres by tape recordings of their calls, and they move about in semi-darkness quite readily. They are, however, very little known and difficult to study. Harriers are much better known and very easy to watch because they hunt almost exclusively in the open. They also have a partial facial ruff and large ear openings sometimes with a decided conch. They hunt upwind, in slow, buoyant, flapping flight, and clearly it would help them to be able to hear a vole moving in dense grass, hover over the sound, and when the prey was exactly located, drop on it. This often appears to be what happens.

We can now leave the head, with its multiple specialized structures and go on down the neck. Vultures often have naked or downy necks, and those with the longest and barest necks are the griffons which feed in masses together on the carcasses of large animals and must often thrust far into quite small openings to reach the soft flesh they seek. No New World vulture has quite such a long, bare neck as a griffon but the very large Andean and California Con-

Head of an Osprey; the eye is not protected by a supra-orbital ridge, and the slit-like nostril in the cere can be closed at will when plunging into water.

dors, which might also feed more on big dead animals, have more bare skin on the head and neck than the smaller Turkey Vulture and Black Vulture, thus paralleling the difference in feathering between griffons and, for instance, the Egyptian Vulture.

The necks of birds of prey are extremely flexible, neither very long nor short, and are not obviously folded in flight except in the long-necked griffons which retract the neck when soaring. Snake eagles, especially, can erect the feathers of the head and neck to form a hood or cowl which may be useful in display; I am tempted to wonder whether this can be used in threatening the sometimes hooded cobras and mambas which snake eagles attack.

In many vultures, the feathers at the base of the neck are more or less adapted for display. The Andean Condor has a snow-white ruff at the base of the neck, and the orange neck of the King Vulture is framed by a collar of filamentous grey feathers. In the Andean Condor, at least, the arched and slightly swollen neck is used in nuptial display. Some large Old World vultures, such as the Lappet-faced Vulture, erect the ruff of long, lanceolate feathers at the base of the neck in threat display to other vultures at a carcass. More information is needed on the ways in which bare necks with bright colours and specialized ruffs are used in display by vultures.

Old World vultures have a large and prominent crop at the base of the neck usually covered with short feathers. The crop of the European Griffon is covered with brown, hair-like feathers encircled with white down, and in vultures generally, it is obviously useful to be able to gorge and store a large meal in the crop. Most other birds of prey have similar if smaller crops but snake eagles can apparently use or dispense with a crop at will. If they eat a rat they store it in the crop but if they eat a snake they adopt the sword-swallower technique and pack it straight down into the intestine.

The bodies of birds of prey are small and light in relation to the area of the wings and the mass of feathers covering them. The size of the breast muscles powering the wings varies a good deal but are often small in relation to the size of the bird. Many vultures and eagles are adapted for effortless soaring and are actually incapable of sustained, flapping flight for long. The breast muscles of a homing pigeon, which must return to its roost by flapping flight over long distances, are huge in proportion to those of a condor or a griffon vulture. Although such vultures can fly a short distance by flapping flight they obviously find it very laborious. Large eagles, too, seldom flap their wings for long; they make a few flaps, glide, perhaps pick up a thermal, and then rise to heights from which they can continue to soar with ease.

The active, bird-killing sparrowhawks and goshawks must be able to put on a sudden burst of speed in flapping flight to catch their prey, and the large, swift falcons which hunt in the open can clearly travel for long distances in steady, flapping flight. In straight flight, however, even the Peregrine cannot outfly a pigeon, or a wader such as a Redshank and must catch such birds with the advantages of height and surprise. A Goshawk, released from the fist can overhaul and catch a powerfully muscled Pheasant if the pursuit does not continue too long. It is a little like the Cheetah which, despite its superior speed, will not pursue a gazelle for more than a few hundred metres because it rapidly becomes exhausted.

The most vigorous flappers of all are the kestrels and the elanine kites which hunt over open country by hovering, continually flapping and twisting their wings through a short arc. Though they reduce the effort by spreading the long tail to obtain extra lift, they must work hard for every mouse they catch. Other hovering birds, notably humming-birds, have very well-developed breast muscles.

Harriers, as a group, probably use flapping flight more continually than any other. Their mode of flapping is a gentle stroking with long wings interspersed with short glides, however, and certainly looks less laborious than the hovering of kestrels. They are among the few birds of prey

Fig. 9 Wings: Common Buzzard – layout of an unspecialized, soaring wing with moderate primary emargination, ten primaries, and ten to eleven secondaries; Bateleur – exceptionally long, 'gliding' wing, with more than twenty secondaries; Black Vulture – soaring wing with strongly emarginated primaries and over twenty secondaries on long humerus; Red Kite – soaring wing with pronounced angle at carpal joint; Hen Harrier – long wing with primaries little emarginated for buoyant, flapping flight; Saker Falcon – long, pointed wing, primaries scarcely emarginated for speed in dive; Shikra – typical accipiter with short, rounded wing, primaries little emarginated, used with long tail for manoeuvring in cover.

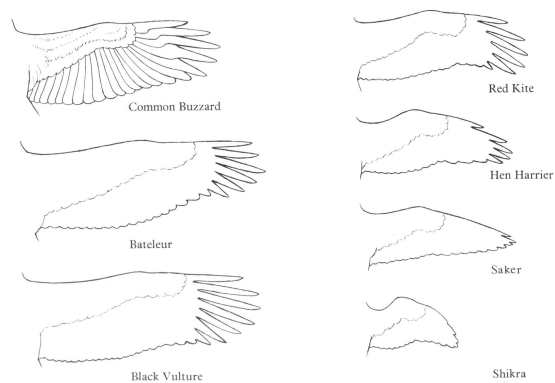

Common Buzzard

Red Kite

Bateleur

Hen Harrier

Saker

Black Vulture

Shikra

Above Sudanese dry savanna after a fire; leafless trees and bare ground look barren but expose possible prey to attack.

The Roadside Hawk, a typical South American savanna buteonine, so called because it often perches beside roads like other buteos.

Left The almost incredible contrast of lush green after good rains is typical of tropical savannas.

that seldom soar but are more ready to cross large bodies of water than most other birds of prey because their flapping flight makes them less dependent on thermals. Harriers have reached Fiji, Reunion, and New Zealand while no typical soaring species has.

The appendages to the body, the wings, tail, legs, and feet, are more specialized for various modes of life than the body itself. In broad terms, wings are for getting about by flight, whether flapping or soaring; tails are for manoeuvring and steering, occasionally with more specialized uses; and feet are for killing and holding prey. The legs are, for the most part, just limbs for perching which are also used to place the feet into position for killing. They are not normally used for walking about, except in the case of the Secretary Bird. Most birds of prey are reluctant and clumsy walkers.

The rather generalized wing of a Common Buzzard is moderately long and moderately broad, adapted for soaring with very limited flapping flight. The layout of the feathers of this typical raptor's wing is displayed in Fig. 9. The wing is composed of strong flight quills attached to the various bones and covered at their bases by the upper and lower lesser, median, and greater wing coverts. The usually ten primaries are attached to the bones of the manus, or hand, the ten secondaries to the bone of the forearm or ulna. Some long tertiary feathers, not specially important in flight, are attached to the base of the humerus. The first digit or thumb is free and is developed into a marked bastard wing or alula, a tuft of stiff, small quills erected or lowered at will to prevent stalling in flight.

The number of primaries (nine to eleven) does not vary much in birds of prey, but their length, shape, and degree of emargination at the tips do. In the Buzzard wing, the five outer primaries are somewhat emarginated on the inner web. When the wing is fully expanded, these emarginations result in notches or slots at the tip of the wing. In vultures, and some big eagles, which are more dependent on soaring than the Buzzard, six or even seven primaries may be emarginated on both webs. This results in deeper, more noticeable wing slots, with square bases. In the wing of a Sparrowhawk, which seldom soars, the primaries are not so deeply or strongly emarginated. Sparrowhawks seldom soar but some big forest eagles, however, with much the same general shape of wing, can soar very well, and long-winged, swift falcons can soar and glide beautifully.

The extreme emargination of the primaries is apparently connected with soaring flight because it also occurs in such diverse birds as pelicans and storks. The effects of the emargination are apparently twofold. Firstly, the wing

slots so formed reduce turbulence at the wingtip, which is useful in landing, and in preventing stalling in flight. Secondly, the widely separated spread feathers can each act as an individual aerofoil or flying surface. It is as if the bird had one big wing and a whole series of little wings attached to its tip. These individual feathers bend to a varying degree under load according to their position and length. The longest primaries, usually numbers 7 and 8, bend more and in level flight lie fractionally one above the other reproducing something of the performance of a biplane or triplane, increasing the lift at the extremity of the wing. This adaptation helps large, soaring birds both to gain extra lift and also to avoid an extremely long wing, most efficient for soaring, like that of an albatross. Very long wings would hinder an eagle trying to catch prey even in light woodland and make landing on the ground or branches difficult; albatrosses are very clumsy on the ground.

The Indian Black Eagle has very long wingtip primaries, which are softer and more flexible than in other eagles of similar size and are very widely spread in the fully extended wing. The African Harrier Hawk has somewhat similar long, soft primaries. I believe that such long, flexible primaries may be an adaptation to slow searching for small, helpless creatures on which both the Black Eagle and the Harrier Hawk depend. In theory, the deep emarginations and great flexibility of the wingtip primaries of the Black Eagle could increase lift and assist slow flight. This

Black Vulture in flight, showing eight spread primaries at the wingtip, square-based wing-slots, bending of the primaries under differential load, and enormously long humerus bearing over twenty secondaries, such very long wings when folded form impressive 'shoulders'.

Above A Laughing Falcon, an aberrant, snake-eating falconid which has the short, rounded wings and long tail characteristic of the forest habitat.

Left The Godare River flowing quietly through tropical forest in south-west Ethiopia. In such places aquatic and forest species of raptors occur together.

Andean Condor

Indian Black Eagle

Common Buzzard

forest falcon (*Micrastur*)

Fig. 10 Emargination of wingtip primaries: Andean Condor – extreme emargination with seven long, stiff primaries emarginated on both webs; Indian Black Eagle – six very long primaries much emarginated, perhaps aiding slow, soaring, searching flight over forest; Common Buzzard – unspecialized, soaring wing with four primaries slightly emarginated; forest falcon – short, rounded wing adapted to forest life but with emarginated primaries.

conjecture needs support by further research, however.

The shape and number of secondaries vary the wing and its performance much more than the differences in shape and emargination of the wingtip primaries. The Buzzard has ten secondaries and a moderately long wing but the Old World vultures have much longer wings with about twenty-five secondaries, and the secondaries themselves are also long so that the wing appears broad and rectangular despite its length. At rest, the long ulna produces the hunched 'shoulders' so conspicuous in big vultures and condors, and which gives big eagles their regal appearance. A goshawk, with a short ulna and fewer secondaries lacks such hunched shoulders.

The unique Bateleur also has an extremely long, specialized wing with about twenty-five secondaries and long, pointed primaries. Because the secondaries are relatively short, however, the extended wing appears long and narrow, and the tips are often swept back to resemble a delta-wing or swept-wing aircraft. This is apparently an adaptation for continuous gliding at a rather high air speed of 55 to 90 kilometres per hour so that a Bateleur covers 300 to 400 kilometres every day. The unrelated American Turkey Vulture also glides much of the day but the almost tailless Bateleur is actually much more specialized. Its short tail is only a fifth of the wing length while the Turkey Vulture's longer tail exceeds half the wing length.

The ratio between the length and breadth of the wing is called the aspect ratio. Long, narrow wings have high aspect ratio while short, broad ones have low aspect ratio. The higher the aspect ratio the more efficient the wing for continuous soaring and gliding. That of a Wandering Albatross is about 20, that of a Fulmar Petrel about 12, but that of a Griffon Vulture about 7. Thus, although you would think of a Griffon Vulture as an efficient soarer, in fact, it has rather poor soaring performance. That of a Golden Eagle should be somewhat better because the relatively short secondaries should improve the aspect ratio; that of a Bateleur, with an equal number of secondaries but a narrower wing still better. The aspect ratio of a Bateleur's wing, however, is actually about 8. Large, soaring vultures and big eagles must compromise between efficient soaring and ability to land and take off easily often in quite small spaces. Both Bateleurs and Lammergeiers have wings of relatively high aspect ratio and both alight quite differently to griffons, with a quick, pigeon-like fanning of their long wings instead of the laboured thump-thump-thump of a big vulture.

Aspect ratio does not vary with weight because it depends on linear measurements. Wing loading, however, which is at least equally important in controlling the performance of any flying machine, bat, or bird, does vary with weight because it depends on the volume or cube of a bird's length. It is the weight carried by any given unit area of the wing, most simply expressed in newtons per square metre (N/m^2). The wing area varies as the square, the volume and weight as the cube, so that the bigger and heavier a bird of a given shape the higher its wing loading. Birds of quite different aspect ratio can have very similar wing loading; for instance, that of the short-winged, forest-loving Crowned Eagle is about 67 N/m^2 and that of the long-winged Golden Eagle of mountains about 65 N/m^2.

Aspect ratio and wing loading can be easily estimated from a live bird or one freshly dead. All that is needed is to weigh the live bird, lay it on its back, and trace the outline of one wing on, preferably, squared paper. Thus, the area, linear dimensions, and weight can be related. We know too little about this subject, and most of the data I have come from the late Colonel R Meinertzhagen. He told me he had sent much more to the Air Ministry, where for all I know it moulders to this day in dusty files marked 'Top Secret'.

All buzzards, from the 700 gram, narrow-winged African Mountain Buzzard to the 1400 gram, broad-winged Augur Buzzard, have remarkably similar wing loading of from 40 to 45 N/m^2; perhaps variations in aspect ratio help to even out wing loading. All known harriers have the lowest wing loading, of 20 to 30 N/m^2, and big eagles and heavy vultures 70 to 120 N/m^2. A 1000 gram harrier would have wing loading of about 30 N/m^2, a similar-sized buzzard about 45, an accipiter about 55, and a swift falcon about 70. Again, the broad wings of accipiters and the long, pointed wings of falcons may have similar total area but quite different aspect ratio.

Broadly, the lower the wing loading the slower a bird can fly and remain airborne, while high wing loading assists high-speed diving. In nature, harriers must fly slowly to use their specialized searching technique while large falcons must be able to dive fast. Accipiters, with wing loadings that vary more sharply related to weight than either buzzards or harriers, never need fly slowly at all.

Some specialized species, such as the Lammergeier, have much lower wing loadings than the Griffon Vulture despite a higher aspect ratio. The Lammergeier has noticeably buoyant flight, and can also glide quite slowly. I forecast, on these principles, that if anyone examined an African Harrier Hawk or an Indian Black Eagle it would prove to have low wing loading. Sure enough, two captive Harrier Hawks kept by J E Cooper and now in the London Zoo, averaged 750 and 950 grams and had wing loadings of 31·5 and 36 N/m^2 respectively, lower than that of a buzzard of equivalent weight.

Wing loading also controls the diameter of the turning

Head of a Lanner Falcon; the toothed beak is used to break the neck of the prey while the round tubercle within the nostril is thought to act as an air-baffle at speed. Falcons normally have dark eyes.

Head of a Monkey-eating Eagle; the reason for the exaggerated deep beak is unknown, but its slimness may aid forward binocular vision; the nostril is long and slit-like, and the head feathers can be erected into a 'fright-mask'. This species was once heavily collected but it is now protected. It is still threatened by destruction of habitat.

Above Head of a buzzard,
showing the large eye protected
by a supra-orbital ridge; bristles
on the lores; yellow cere with
simple nostril; moderately
powerful beak with no teeth,
wide gape, and serrations on the
upper palate which help to
prevent food slipping out.

Left The beautiful coral-red eye
of the Black-shouldered Kite is
shaded and protected against
damage by a strong supra-orbital
ridge of feathers.

Right European Sea Eagle in flapping flight; the seven wingtip primaries bend under load. Such aquatic species often have very broad, deep wings which join the spread short tail to produce maximum flight area, perhaps useful lifting heavy loads.

Far right Fig. 11 Tails: Common Buzzard – unspecialized, moderately long, rounded tail of a soaring hawk; Lammergeier – very long, diamond-shaped tail, use obscure, but perhaps aids slow flight; African Long-tailed Hawk – longest tail of all relative to size; like accipiter, but with central tail feathers long and whole tail graduated; American Swallow-tailed Kite – extreme form of long, forked tail with stiff outer feathers, acting as separate aerofoils for 'biplane' manoeuvrability; Sparrowhawk – long tail used in rapid manoeuvring in dense cover; African Fish Eagle – short tail of a broad-winged, aquatic species; Bateleur – extremely short tail, useless for manoeuvring or steering (which is done with the wings); feet project beyond tip in adult.

circle; the lower the loading the smaller the circle. Thus, a small, soaring bird, such as a Black Kite or Egyptian Vulture, can soar earlier in the day using smaller thermal bubbles than a griffon vulture with much higher wing loading. In the tropics, large vultures cannot normally fly much before 9 a.m., when the increasing heat produces the first big thermal bubbles, and they must normally alight before 5 p.m. Egyptian Vultures, on the other hand, can fly quite soon after dawn, and Black Kites, by combining flapping flight with soaring ability, can search for food at first light. One day in Dessie in Ethiopia, kites were on the wing from 7.25 a.m., almost as soon as they could leave their roosts. Hooded Vultures appeared at 8.30, followed by Lammergeiers at 8.42. The wing loading of a Black Kite is about 25 N/m^2, that of a Hooded Vulture about 48, and that of a Lammergeier about 55. Thus, all these birds appeared just about when they should according to theoretical calculations based on wing loading.

In a steep dive, high wing loading aids speed. Also, the heavier the object relative to its frontal area, the higher the terminal velocity it will reach before air resistance prevents further acceleration. Thus, a Gyrfalcon should both be able to dive faster and attain a higher terminal velocity than the lighter Peregrine because, although it is twice as heavy, its frontal area is not proportionately greater. High wing loading also improves the gliding angle at high speeds and aids cross-country soaring. Adding ballast reduces the rate of sink at high speeds. The blunt-bodied, bullet-shaped, swift falcons are evidently better suited to fast diving than, for instance, harriers or buzzards. Arguments continue about how fast a Peregrine can dive, estimates varying from 38 to 125 metres per second. A Golden Eagle has been timed, on a slightly rising course, at 53 metres per second (by Frazer Darling) and I cannot believe that a Peregrine is limited to 38 metres per second; its theoretical terminal velocity would be about 80 metres per second.

These examples and arguments demonstrate the importance of physical proportions and form of wing, with variations of size, shape, and weight in birds of prey. We need far more accurate data on such subjects, however, before we can draw really sound conclusions. Such data could quite easily be gathered by falconers, zoo keepers, and those who handle migrating birds at ringing stations. These latter would be especially valuable because the birds are wild, and ringers who regularly note such details as weight and wing moult, might also draw the outline of one wing.

Basically, wings are for lifting and propelling the bird and tails are for steering and manoeuvring. Actually, as in a fixed-wing aircraft, where ailerons and rudder are used simultaneously in a tight turn, they have an overlapping complementary function. Tails are more important than wings for steering in confined spaces or forests, and in open country provide comparatively little additional lift even when spread. There is much variation in the proportions of wing and tail, however, often well shown by the length of the inner secondaries. The Augur Buzzard's inner secondaries, for instance, are long so that the spread wing almost joins with the spread tail to form a continuous flying surface when poising or hovering. In Verreaux's Eagle, also a mountain or open-country species, the inner secondaries are very short and there is a big gap between them and the spread tail; this is also true of the long-winged Lammergeier. The precise function of such variations remains to be elucidated by further research.

A Common Buzzard's twelve-feathered tail averages about 54 per cent of wing length which is fairly typical of woodland birds. Closed, it is square ended and fan-shaped when spread in flight. Birds of dense forest have longer, more graduated tails, and those of very open country shorter, square tails. Most accipiters have tails 65 to 85 per

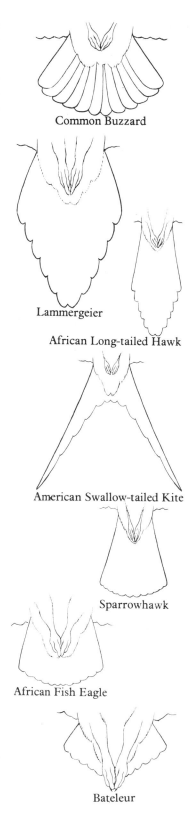

Common Buzzard

Lammergeier

African Long-tailed Hawk

American Swallow-tailed Kite

Sparrowhawk

African Fish Eagle

Bateleur

Right The exaggerated colours and forms of the King Vulture's head and neck are probably nothing to do with carrion feeding but are modified for use in display.

Below The slim, graceful form of a male Hen Harrier (American race) perfectly displayed; long wings with five separated wingtip primaries; long tail, spread and adjusted to brake at landing; long legs useful for grasping prey in thick vegetation, ending in small feet useful for grasping slow-moving mammalian or insect prey. The yellow eye and pale grey colour distinguish the male from the dark-brown female.

Above A Common Buzzard gliding down to kill from a low perch; such hunting requires no very great skill or agility.

Left An accipiter, in this case an adult American Goshawk, about to strike; the feet have been lowered and the red eye is fixed on the victim.

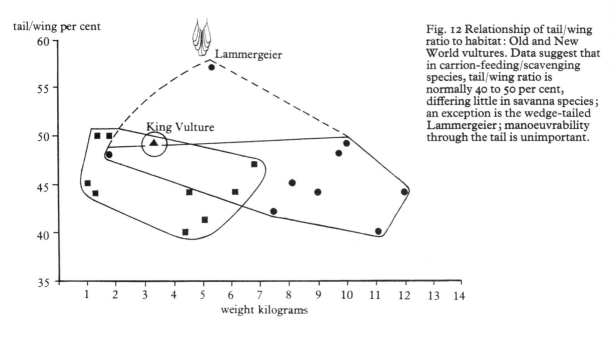

▲ forest spp   ● woodland/savanna spp   ■ open plains or mountain spp

tail/wing per cent

Lammergeier

King Vulture

Fig. 12 Relationship of tail/wing ratio to habitat: Old and New World vultures. Data suggest that in carrion-feeding/scavenging species, tail/wing ratio is normally 40 to 50 per cent, differing little in savanna species; an exception is the wedge-tailed Lammergeier; manoeuvrability through the tail is unimportant.

weight kilograms

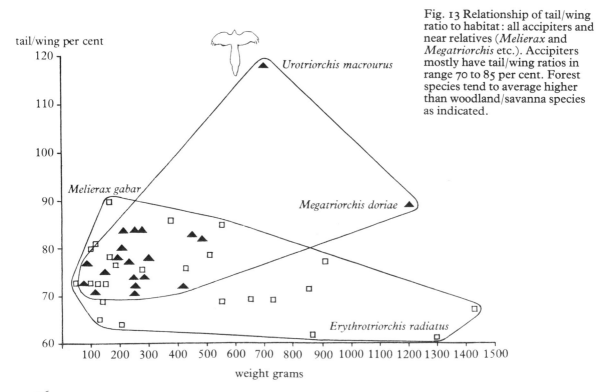

▲ true forest spp   □ woodland and savanna spp, also in forest

tail/wing per cent

Urotriorchis macrourus

Melierax gabar

Megatriorchis doriae ▲

Erythrotriorchis radiatus

Fig. 13 Relationship of tail/wing ratio to habitat: all accipiters and near relatives (*Melierax* and *Megatriorchis* etc.). Accipiters mostly have tail/wing ratios in range 70 to 85 per cent. Forest species tend to average higher than woodland/savanna species as indicated.

weight grams

96

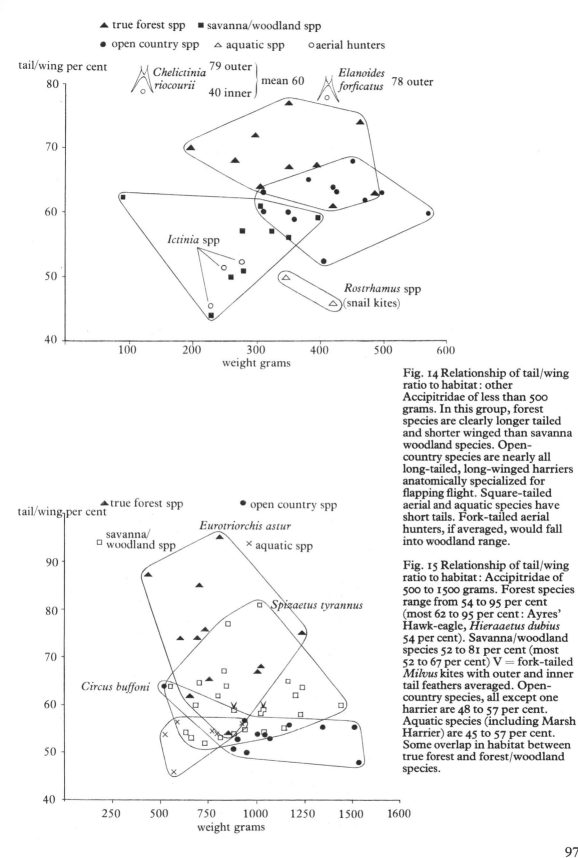

Fig. 14 Relationship of tail/wing ratio to habitat: other Accipitridae of less than 500 grams. In this group, forest species are clearly longer tailed and shorter winged than savanna woodland species. Open-country species are nearly all long-tailed, long-winged harriers anatomically specialized for flapping flight. Square-tailed aerial and aquatic species have short tails. Fork-tailed aerial hunters, if averaged, would fall into woodland range.

Fig. 15 Relationship of tail/wing ratio to habitat: Accipitridae of 500 to 1500 grams. Forest species range from 54 to 95 per cent (most 62 to 95 per cent: Ayres' Hawk-eagle, *Hieraaetus dubius* 54 per cent). Savanna/woodland species 52 to 81 per cent (most 52 to 67 per cent) V = fork-tailed *Milvus* kites with outer and inner tail feathers averaged. Open-country species, all except one harrier are 48 to 57 per cent. Aquatic species (including Marsh Harrier) are 45 to 57 per cent. Some overlap in habitat between true forest and forest/woodland species.

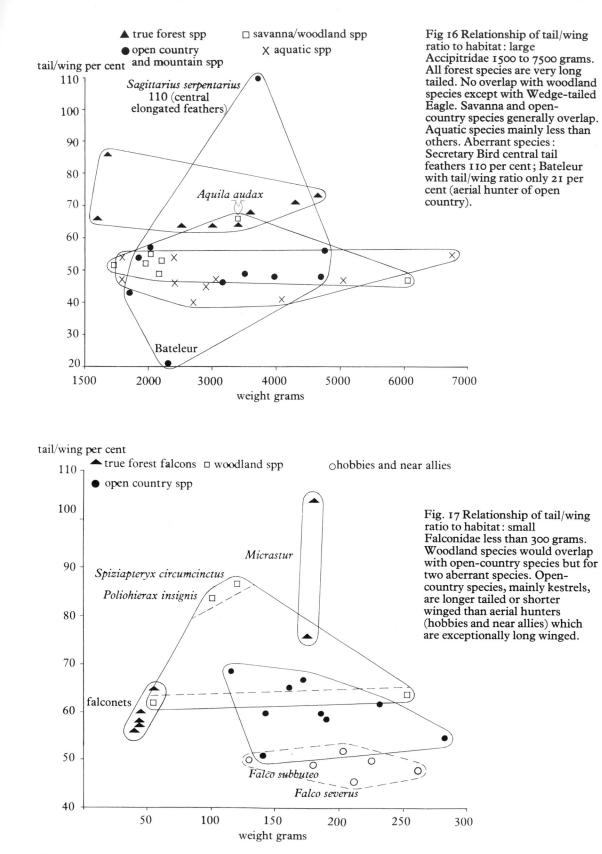

true forest spp ▲   **savanna/woodland spp** □

open country ●
and mountain spp   **aquatic spp** X

tail/wing per cent

Fig 16 Relationship of tail/wing ratio to habitat: large Accipitridae 1500 to 7500 grams. All forest species are very long tailed. No overlap with woodland species except with Wedge-tailed Eagle. Savanna and open-country species generally overlap. Aquatic species mainly less than others. Aberrant species: Secretary Bird central tail feathers 110 per cent; Bateleur with tail/wing ratio only 21 per cent (aerial hunter of open country).

Sagittarius serpentarius 110 (central elongated feathers)

Aquila audax

Bateleur

weight grams

tail/wing per cent

true forest falcons ▲   **woodland spp** □   **hobbies and near allies** ○

open country spp ●

Fig. 17 Relationship of tail/wing ratio to habitat: small Falconidae less than 300 grams. Woodland species would overlap with open-country species but for two aberrant species. Open-country species, mainly kestrels, are longer tailed or shorter winged than aerial hunters (hobbies and near allies) which are exceptionally long winged.

Micrastur

Spiziapteryx circumcinctus

Poliohierax insignis

falconets

Falco subbuteo

Falco severus

weight grams

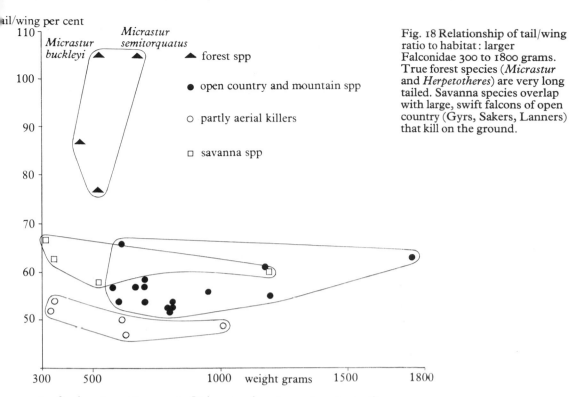

ail/wing per cent

Micrastur buckleyi

Micrastur semitorquatus

▲ forest spp

● open country and mountain spp

○ partly aerial killers

□ savanna spp

weight grams

Fig. 18 Relationship of tail/wing ratio to habitat: larger Falconidae 300 to 1800 grams. True forest species (*Micrastur* and *Herpetotheres*) are very long tailed. Savanna species overlap with large, swift falcons of open country (Gyrs, Sakers, Lanners) that kill on the ground.

cent of wing length, most *Spizaetus* hawk-eagles 65 to 80 per cent of wing length, and forest falcons 75 to 105 per cent of wing length. The graduated tail of the forest-living African Long-tailed Hawk, *Urotriorchis*, is about a fifth longer than the wing. The relationship of wing and tail length to habitat is shown in detail in Figs. 12–18.

The long, fan-shaped spread tails of accipiters and other forest-loving species are used to brake or steer, evidently useful in thick cover. In soaring they are spread to increase the lift obtained from the rather short, inadequate, rounded wings, and in descending from a height such long tails can be raised or lowered to slip air, so accelerating or slowing the rate of descent. In this way, the apparently large and supposedly clumsy Crowned Eagle delicately adjusts its approach to a chosen tree branch, alighting thereon with beautiful precision. The huge Harpy Eagle is said to fly among the big forest trees with extraordinary dexterity by using its long tail. In the forest falcons, which apparently fly very little and never fly high, the long tail is, in cross-section, strongly arched which perhaps helps to reduce abrasion of the long central feathers when running on the ground.

Long tails are not exclusive to woodland or forest species. The outer tail feathers of the fork-tailed American Swallow-tailed Kite, African Swallow-tailed Kite, and the Red Kite are all 65 to 80 per cent of wing length. Averaging outer and inner feathers, however, the proportions become less unusual. A Black Kite manoeuvring adroitly

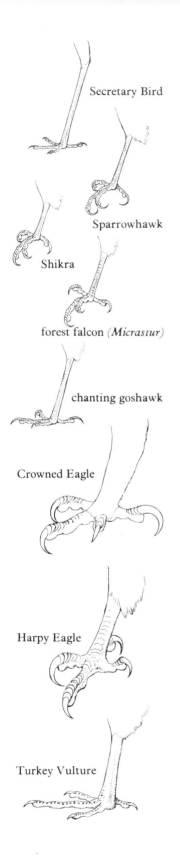

Secretary Bird

Sparrowhawk

Shikra

forest falcon (*Micrastur*)

chanting goshawk

Crowned Eagle

Harpy Eagle

Turkey Vulture

in and out among the telephone wires, pedestrians, traffic, and other hazards to snatch scraps off the streets uses its wings to glide and sometimes to accelerate by a flap or two but twists the tail this way and that to steer, and spreads it suddenly to brake when picking up food. The purpose of the very strongly forked tail of the Swallow-tailed Kites is not entirely clear. When spread, however, they apparently function as a long wing flap, separated by a slot from the trailing edge of the wing, so increasing lift at slow speeds. Alternatively, the spread tail may form a smaller, separate aerofoil lying somewhat below the plane of the wing, giving the bird something of the performance of a biplane. As anyone who, like me, learned to fly on a Tiger Moth knows, a biplane can fly slower than, and out-manoeuvre, a monoplane. Such birds with long, forked tails are specialized for slow flight, hovering, and manoeuvring in confined spaces. The American Swallow-tailed Kite as an extreme form would especially repay aerodynamic study.

A few open-country species, such as the Egyptian Vulture, Lammergeier, and Australian Wedge-tailed Eagle have long, wedge-shaped or graduated tails. The Lammergeier's tail is diamond shaped, almost as long as the body, but is only 60 per cent of the very long wings. Lammergeiers, despite their huge size and wingspan greater than that of the much heavier Griffon Vulture, can glide skilfully in and out between huts. Perhaps the long tail and low wing loading again make for slow-flying ability but the tail is not obviously twisted to manoeuvre like a kite's.

Relatively the shortest tails are those of some fish and sea eagles, though a few are also wedge shaped. They scarcely need to manoeuvre in confined spaces because they catch their prey in open water. The Bateleur has the shortest of all but does not use it for steering. It steers by canting its wings this way and that, hence the name, which is French for balancer or tightrope walker. Tightrope walkers used to carry a long pole which they adjusted to maintain balance.

We can now descend to the ground and consider the legs leading to the feet, the instruments specialized for killing and holding prey.

The legs themselves are seldom used for walking. Only the Secretary Bird really walks kilometres every day. Its exceptionally long legs (about three times as long as those of a similar-sized eagle) end in rather short toes. These are evidently convenient for any bird which must walk long distances through long grass, such as grassland bustards. They are, however, little use for grasping prey in a powerful clutch, and the Secretary Bird tends to kill prey by stamping on it and then swallowing it whole. Its very long central tail feathers, in any short-legged raptor, would simply become worn on the ground. These have nothing to do with flying or steering but are adornments the Secretary Bird can retain because it stands so tall.

Other raptors hunting on the ground have longer legs than usual for birds of their size. Both the European spotted eagles have relatively long legs and both, at times, catch frogs and insects on the ground. The chanting goshawks have relatively long, stout legs, long enough when standing to lift their long tails completely clear of the ground. The crab hawks (*Buteogallus*) also have longer legs than buteos of comparable size; these would evidently be useful when hunting crabs in mud and shallow water.

Forest falcons also have relatively long legs, and though very little is known about them in the wild state, the behaviour of a captive bird showed that they can run freely about on the ground, along branches, or through undergrowth. This captive pursued a rabbit on foot and caught it in cover too dense to fly through, behaving like a Roadrunner (*Geococcyx*). It also walked about on the ground and from there leaped or flew to higher branches to catch arboreal mammals. All forest falcons, therefore, probably run about on their long legs but also need long tails for manoeuvring in thick undergrowth.

Long legs are otherwise mainly used to extend the reach and grasp when killing. Some sparrowhawks have long, thin shanks with long, thin toes and needle-sharp talons. Evidently, this gives a bird killer extra 'reach' while the wide foot gives better grasping power. The long legs of harriers are also for grasping prey deep down in rank vegetation but their toes are shorter and stubbier because they kill mainly small mammals and do not need such a wide spread of foot. Similarly, Ayres' Hawk-eagle, feeding mainly on small birds, has very long legs for its size. The very large and powerful, gamebird-eating Martial Eagle has relatively much longer legs and a wider spread of rather less powerful toes than the slightly smaller, related Crowned Eagle which feeds mainly on large mammals. It needs rather shorter legs with thick, enormously powerful toes and talons to subdue strong, sometimes violently struggling prey up to four times its own weight.

The relative proportions of feet, legs, bill, and wings have been studied in great detail in accipiters by J Wattel. Although he concludes that no absolute separation into different types is possible because of intermediate forms, he considers that long legs, with needle-sharp talons and long toes suggest habitual bird killing in flight while short legs, with relatively short, thick toes suggest larger, less agile prey taken on the ground. A very long middle toe also suggests aerial bird killing and a powerful hind toe with a strong talon suggests ability to kill quite large mammals.

The genus *Accipiter* was originally divided into two,

Far left Fig. 19 Feet: Secretary Bird – very long leg with short toes for walking in grass; Sparrowhawk – long, bird-catching leg with long, slender toes; Shikra – shorter, thicker leg with stronger toes, lizard-eating; forest falcon – long, bird-catching leg resembling that of Sparrowhawk; chanting goshawk – relatively very long leg assists hunting on ground; Crowned Eagle – very powerful, large, mammal-killing leg with enormous hind talon; Harpy Eagle – very powerful, mammal-killing leg, with bare tarsus; Turkey Vulture – chicken-like foot with short, blunt claws, rudimentary hind toe.

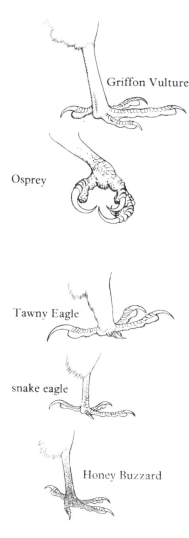

Above Fig. 19 *cont.* Griffon Vulture – carrion feeding means that powerful claws are unnecessary and have been lost; Osprey – foot with spicules, long claw; outer toe is reversible; Tawny Eagle – moderately powerful foot suited to killing smaller mammals; hind claw relatively short and weak; snake eagle – short, stubby, powerful toes suited to grasping and killing snakes; Honey Buzzard – toes with talons blunted by digging for wasps' nests.

Accipiter with long legs and thin toes, and *Astur* with shorter legs and stronger toes, typified by the European Sparrowhawk and Goshawk respectively. Behaviour partially supports this original division because the Sparrowhawk is almost exclusively a small bird killer, while the Goshawk often kills rabbits and occasionally a hare. The Goshawk too, is, however, *mainly* a bird killer. The small Shikra, *Accipiter (Astur) badius*, has short tarsi and short toes adapted to ground predation, and is said to live mainly on lizards. The Levant Sparrowhawk and the Grey Frog Hawk, the latter mainly eating frogs, have similar legs and feet. A parallel can be drawn among small falcons with the Hobby and the Kestrel, the Hobby hunting birds or insects in the air and having long legs and thin toes, and the shorter-legged, thicker-toed Kestrel mainly catching voles on the ground.

While such measurements in little-known species can give useful clues to possible preferences, they must be supported by adequate lists of prey taken, which are often very difficult to obtain, particularly in tropical forest. A study by A Brosset of four African forest accipiters (the large Black Sparrowhawk, the medium-sized African Goshawk, the small Chestnut-bellied Sparrowhawk, and the tiny Rufous-thighed Sparrowhawk) suggests that they kill different types of prey and should not, therefore, compete appreciably. It is not supported by lists of prey taken, however, though the estimate from the structure of its feet and bill that the Black Sparrowhawk takes mainly birds is correct.

The feet themselves are not always specialized for killing. The toes of New and Old World vultures end in short, blunt claws and in the New World vultures the hind toe is rudimentary; their feet resemble those of chickens. Some Old World vultures, such as the Lammergeier may still be partly predatory and may kill some live animals, while the Lappet-faced and White-headed Vultures almost certainly do kill gazelle calves or hares. The crashing impact of their heavy bodies may be enough to disable such prey, whereas a lighter eagle would kill such animals by sheer gripping strength after impact. Vultures probably once had more powerful killing feet but the carrion-eating habit has made strong claws and talons unnecessary so that they have been lost, while hooked bills for tearing flesh are retained.

Species which kill particular prey share similar specialized adaptations of the feet. The Osprey, feeding entirely on fishes, has the most specialized fish-killing foot. It alone has a reversible outer toe and the soles of all its toes are armed with sharp spicules which help to grip the slippery fish, while the reversible outer toe is useful for carrying the prey. Fish and sea eagles, which are less exclusively dependent on fishes than the Osprey, have spicules on their feet but no reversible toe; their way of killing fishes is also different. The only other mainly fish-eating species is the South American Fishing Buzzard. It also has spicules on the feet and interestingly has also developed a mainly white head and chestnut plumage, rather resembling some Old World fish eagles and the Brahminy Kite.

Crab hawks have rather long legs and rather short, strong toes, doubtless useful for crushing and subduing crabs. The Savanna Hawk, also has long legs and rather short toes and feeds much on water snakes and frogs in marshes. The snake eagles of the genera *Circaetus* and *Spilornis* all have rather short, strong legs covered with large rough scales and ending in short, very powerful toes. You can appreciate the usefulness of such toes in subduing a wriggling, possibly venomous snake by picking up a piece of plastic hose and passing it through your fingers; evidently short, stubby fingers rather than long 'pianist's' fingers would give a better grip. The Bateleur, less dependent on snakes, but still killing some, retains the typical snake eagle foot though its wings and tail are specialized. The unrelated South American Laughing Falcon, which also kills snakes, likewise has rather short legs and powerful, thick toes. It apparently kills snakes at least partly by a heavy impact and then uses its toothed, falcon-like bill to bite off the head or break the neck.

The claws of the Honey Buzzard are often blunted by digging like a terrier to reach underground wasps' nests. The Indian Black Eagle, a slow, soaring searcher apparently feeding mainly on young birds and eggs, has very long, rather gently curved talons which would evidently give its foot an unusually wide spread and would help to snatch up a whole bird's nest without pausing in flight. In this species, however, we know too little of what actually happens.

Most of the bigger birds of prey which feed on large animals kill by the sheer force of their grip, which must be experienced to be believed. Big, mammal-eating eagles have very thick, strong toes and the hind talon is long, thick, and sharply curved. In practice, the bird crashes down on its prey and grips it with three toes pointing forward and one back. The huge, dagger-like hind talon can then be driven right into the body while the grip is suffocating, if the impact has not already disabled the prey. Despite the enormous strength and apparent insensitivity of such an eagle's foot, I believe it can be delicately positioned when subduing very large, struggling prey until the hind talon can pierce a vital spot.

A long, very strong, hind claw alone indicates that an eagle kills larger animals than would be expected. That of Bonelli's Eagle is very long and strong for its size and that

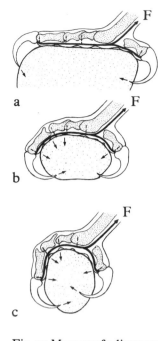

Fig. 20 Manner of adjustment of the foot to differently sized prey (after G E Goslow):
(a) with large prey, the force is principally exerted at the talon-tip;
(b) with moderately sized prey, outer parts of the toes also exert crushing force;
(c) with small prey, crushing grip of toes and foot becomes more important than talons.
F = force of flex on muscles.
Reproduced by courtesy of Auk *magazine.*

of a Golden or a Verreaux's Eagle is relatively longer and stronger than that of a Tawny or Steppe Eagle. Such comparisons cannot be carried too far, however; recent research has shown that Martial Eagles, Tawny Eagles, and Bateleurs, with very different forms of feet and hind claws all feed largely on dik-dik in the Tsavo National Park.

The way in which the foot is used to grip and subdue prey has been carefully studied by G E Goslow in some American raptors. The foot can adjust to objects of different size and, as the talons close, the force exerted at the end of the hind or other claws is increased, while pressure is also exerted by the inside of the toes themselves. The toes are controlled by two powerful muscles acting together, the *flexor digitorum longus* which contracts the fore toes and the *flexor hallucis longus* which contracts the hind toe with its talon. Whatever the size of the object seized, the contraction of these muscles closes the foot until it meets resistance. A snake eagle, with its short toes, would have difficulty in wrapping its feet round a rabbit and would seldom attack it, while a very long-toed, bird-killing accipiter would have difficulty in subduing a snake by closing its toes. Another muscle, the *tibialis anterior*, flexes the leg joint, which is undoubtedly useful at the moment of strike, when the raptor may hold the prey close to the body. Very long legs, such as those of the Secretary Bird, mean little grip in the feet. Such details are better illustrated by diagrams than by descriptions which are apt to become too technical.

In large falcons which strike large birds in flight with the hind claw, the *flexor hallucis longus* is important and the legs are thickened to withstand the heavy impact. The hind claws of larger female falcons are bigger and stronger than those of the smaller, lighter males, suggesting that they can kill bigger birds. In kestrels, which do not kill by striking in the air, but grip mammalian prey on the ground, these adaptations are less evident.

Anyone examining any bird of prey, working with the available knowledge on the functions of specialized structures, could make a fair guess as to where it lived, how fast it could fly, and what it ate by examining its bill, wings, tail, legs, and feet. Supplementary field studies are always necessary to show precisely how these various structures are used and conjectures as to the preferred type of prey must be substantiated by detailed facts carefully and laboriously gathered in the field. Wild birds have a way of being more adaptable and variable in behaviour than their appearance would lead us to think. I have seen supposedly sluggish, lizard-eating chanting goshawks kill full-grown guineafowl and quail in full flight, but from their structure you would hardly conclude that they could perform such feats at all.

Hunting and feeding methods

Considering that the way in which birds of prey kill is perhaps the most spectacular and interesting aspect of their lives, it is surprising how little we really know about how it is done. It is one thing to deduce from the structure of wings, tail, feet, and bill in a museum skin how that bird obtains its food; it is quite another to see it happen often enough to draw sound conclusions. Most experienced observers of birds of prey admit they have seldom seen their subjects kill even by accident. In thousands of hours of watching Verreaux's Eagles in the Matopos Hills of Rhodesia observers have only seen the Eagles kill hyrax twice – both times in a few seconds. In thirty-five years of intermittent watching, I have seen Golden Eagles kill five times. Although I see the Crowned Eagle in the forest opposite my home in Kenya almost every day, I have never seen them kill, but I have several times missed the action by a few moments and been able to reconstruct the event. In such thick forest I could only be in the right spot at the right time by pure chance. Thus, we can say of many birds of prey, 'This and that should happen'. But does it?

For many species the methods can be pieced together from scattered observations. We can improve on this piecemeal, scattered approach by systematic, careful watching of particular species for long periods. Thus, I myself have in recent years watched fish eagles, usually two pairs at a time, for continuous periods of up to 48 hours. As a result, I can now say with fair certainty how often and how they kill and how much they eat per day. Many years ago, Gustav Rudebeck watched raptors on the Swedish Coast intensively and wrote a pioneer paper about their hunting success. A few others have tried the same but sometimes come to quite different conclusions. In many species, however, we can still only guess how the bird catches its food.

The difficulty of obtaining good, quantitative data on this subject depends on, firstly, the density of the habitat, and secondly, on the activity and wide-ranging powers of the bird of prey itself. Secretive accipiters are obviously very hard to watch in dense forest and it becomes almost hopeless to follow a Bateleur through roadless African savannas, or even a Golden Eagle in the Highlands of Scotland. A third factor is the frequency of killing and the type of prey taken; evidently, a buzzard which kills six rats in open grassland per day is more likely to be seen killing than a Crowned Eagle which kills one antelope inside forest about every three days.

Possibly, such forest-loving species can best be studied by catching them, fitting them with little radio transmitters (very small ones are now available), and then following them about on foot with a portable receiver. Even then, the dense forest interferes with the radio signal so that it cannot be located at long range. If the bird is approached too closely, it may not behave quite normally, but if we try just to locate it and then keep out of sight, we will miss the actual moment of a kill and important details. Still, any such results would be better than none.

It is nearly as difficult to follow what happens in a large, wide-ranging species, such as a Golden Eagle, which may daily cover many square kilometres. Such birds usually kill in the open, however, and we have a better chance of actually seeing a kill. Here again, radio transmitters have so far proved of limited value. If they are small enough not to hamper the bird, their signal is inaudible at more than a kilometre or two, or when a mountain blocks the signal as the bird flies into the next valley. Great improvements have been made in these gadgets lately and I believe that much work done remains unpublished.

The unspecialized Common Buzzard in Britain normally inhabits no more than 3 square kilometres, hunts most of its prey in the open, either from perches or by hovering, and is comparatively easy to watch. It, and several other similar buteos, have been quite intensively observed. The Craigheads in Michigan followed two species of buteos, some Hen Harriers, and American Kestrels (Sparrowhawks) from dawn to dark, and from these observations drew some general conclusions on hunting behaviour. On Dartmoor, Peter Dare concentrated on Buzzards only, and watched them the whole year round. Among buteos, there is quite good correlation between scattered observations and detailed observations as regards general hunting methods and prey taken. Scattered observations can never tell us how much food they ate.

The amount needed to sustain life can be estimated from exercised captive birds. Being captives they perhaps do not get as much exercise and eat less than wild birds. Again, however, this approach is by far the best we have and once more the Craigheads recorded quite invaluable facts on the amounts eaten by various species, from Golden Eagles to Kestrels, winter and summer, and with and without exercise. Unfortunately, few such facts have been recorded since, although hundreds of birds of prey are kept captive in zoos by falconers or people who just want to keep a bird of prey. The measured needs of captives are hard to relate to the consumption of wild birds of the same species because we can seldom weigh what the wild bird catches. It can sometimes be estimated from prey preferences, and twice I have actually weighed animals eaten by Crowned Eagles.

The easiest of all situations for systematic watching is the waterside. Ospreys, sea and fish eagles, and several other species such as Eleonora's Falcon hunt only or mainly over open water. In these cases, long enough watches can reveal how often a kill is made, its identity, and often its weight. At Lake Naivasha I have not only obtained quite good correlation between scattered daily observations and detailed dawn to dark watches, but can say with fair certainty that a Fish Eagle kills about 130 to 140 grams of fish per day, and eats nearly all of it, bones and all, with little waste. This nicely agrees with a theoretical forecast of food needs based upon body weight derived from the Craigheads' figures for captive birds.

Likewise, those who have observed the Osprey find it relatively easy to record what it catches, how long it takes to catch its daily needs, and its hunting success (often very high). I know of no such accurate data for the American Bald Eagle or the European Sea Eagle but similar facts should not be too difficult to gather in a well-chosen situation such as the skerries off the Norwegian coast. In such cases, small radio receivers carried in boats could help locate a bird temporarily lost.

In the United States, small radio transmitters have been used successfully on Bald Eagles to determine the pattern of activity in winter. The various signals showed whether the Eagle was perching, flapping, soaring, and so on but although the results generally confirmed sight observations, they did not give quantitative facts about food consumed or prey taken. Likewise, small transmitters placed inside dead fish have shown that Ospreys in Minnesota flew an average of 2·6 kilometres per fishing flight, and that they were selective, preferring Bluegill Sunfish and crappies to larger suckers and pike.

The methods used by raptors to catch their prey are many and varied, and I shall survey them by starting with insects and going up the size scale to elephants and whales as carrion. Species which feed on insects either in the air or on the ground are almost as easy to watch as aquatic raptors, though if a kestrel kills in long grass we often cannot be certain of the result. Such largely aerial feeders as the American Swallow-tailed and Mississippi Kites, European Hobbies, or Black Kites, however, can be watched catching flying insects, normally in the foot, then biting off and dropping the thorax, swallowing the edible abdomen. In African rainy seasons, an emerging swarm of flying termites may be gobbled on the ground by a Lappet-faced or Egyptian Vulture, a Tawny or a Ver-

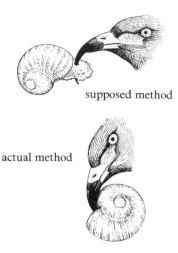

supposed method

actual method

Fig. 21 The Snail Kite's method of extracting the snail. It was long supposed to pierce the snail on its bill and wait for it to relax, but it actually inserts its long, curved spike round the helix of the shell to cut the columellar muscle (and remove the animal in bits).

reaux's Eagle, an Augur Buzzard or a chanting goshawk, while the insects that manage to escape these and fly may be taken by the Black Kites, Hobbies, and Lanner Falcons. Systematic watching will show how many termites a Tawny Eagle eats on the ground or, more difficult, a Hobby in the air.

We would seldom see all such species together but I have often seen four or five raptor species feeding on a termite swarm. Hobbies, Red-footed Falcons, and even large Steppe Eagles make a staple diet of termites in winter in southern Africa, though in Kenya and Ethiopia, Steppe Eagles feed mainly on rodents. Many animals, man included, will eat termites when they can. Locusts used also to be a standby but in recent years have been largely controlled. Although such insects are fat and nutritious, evidently a Steppe Eagle needs a very large number of termites for a full meal. A Secretary Bird or a kestrel hunting large grasshoppers evidently needs fewer, but still a good many. Again, we can watch such birds, then weigh a hundred grasshoppers or termites to arrive at a quantitative estimate of food taken.

The Honey Buzzard, which is not a buzzard and only eats honey incidentally, is seeking the wasp grubs in bulk rather than individual insects. Perching at the edge of an open space, it watches keenly for wasps returning to their underground nests. It then digs at the spot, often completely disappearing and eventually, it digs up a mass of papery comb filled with nutritious grubs. It eats a few adult wasps incidentally, seizing them by the middle, and nipping off the sting before swallowing them. In Africa it snatches the entire nest of arboreal wasps and hornets from beneath the eaves of houses. The American Caracara, another wasp specialist, hangs upside down from, and systematically eats the grubs from the papery nests of tropical American hornets.

Other sluggish, poor-sighted creatures such as snails, slugs, earthworms, and perhaps frogs which may be caught on the ground may be easier to catch than insects. Kestrels skin slugs before swallowing them. Snails are especially interesting but are only eaten in numbers by the South American Snail Kite and the Hook-billed Kite. According to a highly circumstantial and convincing account by Lang, in Guyana, the Snail Kite picked up the *Pomacea* snail, flew to a perch, and waited till the animal emerged of its own volition and tried to crawl away. Like a flash, the Kite then pierced the body of the snail on its needle-sharp bill, immediately pushed it up the mandible till it stood out like a walnut, waited till the snail relaxed, then shook its head so that the empty shell flew off, and swallowed the snail 'operculum and all'.

This account, apparently too convincing to be untrue,

has been repeated from book to book and has long obscured the true facts. Some critics doubted the possibility of the snail being shaken so easily free of its shell because the animal is attached by a strong columellar muscle to the shell, and even if it (improbably) relaxed through being pierced at a nerve ganglion it could not just be shaken free.

In fact, as recent observations in Florida and elsewhere have shown, the Kite picks up the snail in its foot, and then perches and adjusts its position so that it can insert the point of the bill between the operculum and the shell to remove and discard the operculum. Readjusting the snail's position, the bill is now pushed round the coil of the shell till the columellar muscle is cut, and the snail can then be extracted easily. The soft body of the snail is eaten at a gulp or in several bites. A metal tool can exactly reproduce the action of the Kite's pointed bill. Yet, I hesitate to reject as a plain untruth so circumstantial and neat an account as Lang's. The *Pomacea* snails concerned are large and round, and possibly few other raptors feed thus on snails because the molluscs elsewhere are less conveniently shaped. The common giant land snails of Africa, *Achatina*, should be easily caught and nutritious.

The African Harrier Hawk and South American Crane Hawk which feed on rather small, helpless creatures found by methodical searching, have double-jointed legs uniquely adapted to probe crannies. The Harrier Hawk can bend its leg about 30 degrees beyond the vertical so that it can hang on to the edge of a hole in a tree with one foot and push the other deep into the cavity, perhaps to catch a bat or a fledgling barbet. A Crane Hawk, searching a tree with

Above An African Harrier Hawk, a specialist hunter with double-jointed legs which it can thrust into crannies, meticulously searches a dead tree for small prey.

Fig. 22 The double-jointed leg of the Harrier Hawk enables it to reach inside a cavity to snatch nestlings from the bottom or a roosting bat from the top.

Fig. 23 A Rough-legged Buzzard descends from a hover to catch a vole; nearing the ground, the wings are raised and the legs lowered in a final pounce after a cautious approach.

large flakes of bark splitting off the trunk, pushes its head and beak into the larger openings and its double-jointed foot into narrower openings. Both these birds can reach prey accessible to no other raptor and both spend long periods methodically searching places where such prey could be found. The Harrier Hawk, too, hangs from, and systematically destroys weaver birds' nests to take the nestlings. I have already mentioned that its flexible primaries and low wing loading should adapt it for slow, searching flight.

Frogs and amphibians are eaten by harriers, snake eagles, fish eagles, some buteos and kites, even by one accipiter (the Grey Frog Hawk), but seldom as a staple, usually as a sideline. Even the Golden Eagle does not disdain frogs. They are most easily caught when spawning when they swarm in shallow pools. Lesser and Greater Spotted Eagles hunt small frogs on foot in marshy meadows, and in Hungary the Red-footed Falcon eats a great many. Frogs and toads are rather helpless creatures; their only real defence is to hide underwater or in vegetation though they sometimes exude poisonous fluids from their skin. They are not very important as raptor diet, and their remains are difficult to detect in castings because the bones are wholly digested.

Rats and mice are a staple diet for many species, from kestrels and elanine kites to the Steppe Eagle and Saker Falcon. Many raptors eat them when they are abundant; several, such as kestrels, elanine kites, harriers, buteos, and some small eagles feed chiefly on rats and mice varied by grasshoppers, small ground birds, and other easy prey. Rats and mice are caught either from perches or from flight, often nowadays from fence or telegraph poles, or electricity pylons and wires. The perched raptor watches the ground intently. On seeing a movement in the grass it leans forward, often moves its head to and fro, and its gaze sharpens. The quarry precisely located, it launches itself in a short glide, or descends like a parachute, wings upheld, accelerating sharply at the last minute to drop nearly vertically with wings raised on the prey. The impact of landing bends the legs, and automatically contracts the foot. An 80-gram rodent is almost immediately crushed or paralysed by an 800-gram buzzard, or even a 200-gram kestrel. Yet one such easily caught rat will more than suffice the kestrel for a day, and two or three at most, the buzzard.

In steppes and open grass plains, lacking perches, soaring, hovering, or coursing (harriers) must substitute for them. Most raptors hunting open grasslands hover, from kestrels to buteos, even snake eagles. In a good head wind, actual fanning of the wings may not be needed, but in a light breeze the bird must maintain position by rapidly fanning its wings and expanding the tail to obtain extra lift. A slow-motion film of any hovering bird shows that the head remains fixed in one position while the body, wings, and tail shift about. The hover is, in effect, an aerial perch and when prey is finally located it is killed in much the same way as from a post or tree. The hovering bird descends slowly, perhaps checks once or twice, then drops suddenly in the grass with wings upheld. If it rises at once, it probably missed but may have caught something small, but if it stays down the prey can be identified.

When hunting small mammals over grassland, harriers fly head to wind in slow, flapping flight working from side to side, often travelling over the ground at scarcely more than a walking pace. The advantage of slow movement be-

comes evident if you consider looking for a lost golf ball by sprinting or by walking slowly. A hunting harrier periodically checks, briefly hovers in flight, and may apparently follow something it sees, or possibly hears, for a few metres, finally dropping on to its prey in rank vegetation. It may be a grasshopper, a frog, or a nestling bird, but often is a small mammal.

Some swift falcons use a totally different method to catch rats. The Saker is said to hover over Russian steppes like a gigantic kestrel seeking hamsters but also sweeps at great speed a few centimetres above the open grassland, trying to surprise a rat. The Lanner uses similar methods in highland Ethiopia, surprising rats standing at the mouths of their burrows.

Some larger, burrowing mammals, such as gophers, moles, or mole rats are still easier to catch. An Augur Buzzard or Tawny Eagle can just sit on the ground near where a mole rat (*Tachyoryctes*) is working close to the surface. The mole rat cannot see the predator at all but if it approaches the surface, the bird need only fly silently a metre or so and grab. Its closing talons will penetrate the loose soil to grip the rat, pulling it out and crushing it.

Common Buzzards eat many moles, and Tawny and

A hovering Kestrel; the lift from the gently fanning wings is augmented by the spread tail, and the searching head turns this way and that.

Steppe Eagles and the Augur Buzzard, which makes a staple of mole rats in the Kenya Rift Valley cannot have any difficulty in obtaining their daily needs because the large rodents often swarm. Other larger rodents, such as the American Ground Squirrel (a staple of the Alaskan Golden Eagle) are much more alert and less easily caught. They must be surprised in the open before they can escape in a burrow and this may need a planned, long-distance attack. In Utah I once saw a Golden Eagle, flying fast and low over the ground, seize a Ground Squirrel behind a sage bush. It had probably seen the Squirrel from far away, and the swift, low attack was intentional.

Most of the prey so far discussed has been helpless or

Ground Squirrels at the mouth of their burrow; although apparently obvious and vulnerable, they are very alert and not often caught.

sluggish or both. Snakes may be neither, and venomous species such as mambas, cobras, or Puff Adders can be dangerous. Snakes are hunted by snake eagles in the Old World and by the Laughing Falcon and probably by the little-known forest hawks of the genus *Leucopternis* in America. Buteos and eagles also take some snakes, including venomous rattlesnakes and Puff Adders but again as a sideline, not a staple. Snake eagles, on the other hand, are not excited by anything but snakes. A hungry, captive snake eagle will disinterestedly regard a rat but wriggle a snaky-looking bit of rope and it wakes up at once.

Precisely how snake eagles kill their prey is not yet clearly known but probably they drop on it from a perch or a hover with a thump, possibly hitting it near the head.

Snakes have very fragile, easily broken backbones and are disabled behind such a break. The short, thick toes of snake eagles are adapted to grasping a thin, wriggling snake, and a small one may be snatched into the air almost at once where it hangs largely helpless. It is then passed through the feet till the head is reached, which is then mangled with the beak.

A large cobra, or a stout-bodied Puff Adder presents more of a problem and uch encounters must sometimes be epic. The rough scales on snake eagles' legs are supposed to protect them from snake bites but apparently snakes strike mainly at the body and wings of the eagle, to bite only a mouthful of feathers, so that none of the venom enters the eagle's bloodstream. In due course, the snake is exhausted, its back broken, and the potentially dangerous head is eventually crushed. A big snake is then torn in pieces, a smaller one swallowed whole, packed in, hand over hand as it were, till all but the tip of the tail has disappeared, often while the eagle is in flight. The Laughing Falcon is reported to swallow snakes tail first but all true snake eagles swallow their prey head first. The venom is digested and cannot harm the eagle unless it has an internal wound through which the venom could enter the bloodstream.

A highly versatile predator such as a buzzard can deal with prey from venomous snakes to rodents and frogs and insects but seldom snails or fish. The preferred prey of the Common Buzzard, however, throughout its range is small and medium-sized mammals, such as young rabbits. Rabbits are, or were before myxomatosis, almost vital to British Buzzards but in Europe various other small rodents are much more important. Several other buteos, such as the American Red-tailed Hawk and the Rough-legged Buzzard, are mainly dependent on small rodents. Buzzards can adapt their hunting and feeding methods to the available prey and the environment; in the New Forest, British Buzzards hunt more birds in the manner of an accipiter, and the American Red-tailed Hawk kills many Pheasants.

Most buzzards or buteos hunt from perches, either within or outside cover, frequently on a prominent dead limb. They also soar and hover in open country, suiting the method to the prey available and the conditions. In winter in Michigan, the Craigheads found that the diet of three species of buteos (Red-tailed, Red-shouldered, and Rough-legged) was composed of more than 90 per cent small rodents, caught mostly from perches. On Dartmoor, in years just after myxomatosis when the rabbit population was very scarce, the Common Buzzard hunted mostly from perches between September and February, but from March to August mainly hunted by hovering over open ground, using perches only when bad or too calm weather prevented hovering. Perches near rabbit warrens were also used for the specific purpose of hunting rabbits. The Buzzard also walked about on the ground hunting beetles mainly in autumn and winter, not in summer, and if a breeze arose during a calm summer day would often start to hover. In this case, hovering was apparently associated with a shortage of small mammals, but this would not be true of all such cases because many buteos habitually hover when small mammals are abundant.

The British Common Buzzard feeds mainly when it can on the largest mammals easily caught, such as young rabbits. In Australia, since the introduction of rabbits, these have become the staple prey of most of the larger raptors, and in Hawaii the sole raptor present, a buteo, now feeds mainly on introduced rats whereas it must originally have eaten other prey. A versatile predator such as a buzzard will utilize whatever is most abundantly available, although if plenty of a wide variety of prey is available, it may concentrate on preferred items. Specialized predators, however, must stick to their speciality. The Snail Kite and snake eagles would starve in areas swarming with rabbits, though a snake eagle might adapt better than a Snail Kite.

Birds are caught in a variety of ways, starting as eggs when they may be snatched up by the Indian Black Eagle or American Swallow-tailed Kite without pausing in flight; harriers also eat some birds' eggs.

The Egyptian Vulture may pick up whole pelican and flamingo eggs in colonies and break them by hurling them against the ground or rocks. The Egyptian Vulture also hurls stones against Ostrich eggs to break them, thus joining the select band of tool users (including man, higher apes, the Sea Otter, the Galápagos Woodpecker Finch, and possibly a bower bird). Emu eggs are likewise said by Australian Aborigines to be smashed by the Black-breasted Buzzard Kite, by dropping stones on the nests, first having driven off the Emu by a threat display. This remains unproven, however.

Harriers take many young birds, and fledglings just out of the nest are easy prey to accipiters. Welsh Red Kites have, in recent years, been shown to eat many young corvids and gulls, and to be more agile bird eaters than might be supposed. Young water birds also are often taken out of the nest by some sea and fish eagles; I have watched an African Fish Eagle decimate a colony of African Spoonbills, going from nest to nest till most were eaten. Such young birds present no great challenge to hunting skill but sea and fish eagles also take adult diving ducks, auks, and other water birds.

Adult birds are caught mainly by bird-hunting accipiters, by the swift falcons, and by some large and small eagles. The size and type varies according to the size of the raptor. A list of about 95,000 items shows that 98·95 per cent of the Sparrowhawk's prey by weight is birds, 85·1 per cent by weight is small passerines of 10 to 50 grams, one fifteenth to one quarter of the Sparrowhawk's own weight. In Holland, House Sparrows were the commonest but this varies from place to place according to availability. In about 13 000 Goshawk kills 90·2 per cent by number and 89·8 per cent by weight were birds (showing that, despite its powerful feet, this accipiter too is mainly a bird eater) but here only 28 per cent by number and 6·3 per cent by weight were small passerines. The commonest birds taken were non-passerines (chiefly pigeons), 48·3 per cent by weight. The goshawk could take anything from a

A cock Pheasant, brilliantly coloured and vulnerable in the open, can protect himself only by alertness.

cock Pheasant to a thrush but preferred Woodpigeons, while 9·8 per cent of its prey by weight was rabbits and hares.

Although an accipiter kill is seldom seen because it usually happens in or behind thick cover, kills are by surprise and dexterity rather than sheer speed. The hawk flies along a line of trees or a hedge and flicks over the top suddenly to surprise a bird feeding in the open. Alternatively, the hawk approaches low over the ground, behind a bush or other cover, and suddenly dashes among a flock of feeding sparrows. Such attacks may have been planned some distance away. Hawks may also hunt from perches within cover, snatch a bird flying past, or suddenly arrive in a tree and snatch a feeding bird.

Accipiters are selective hunters seldom catching birds which habitually skulk in small numbers in dense thickets,

This female Goshawk has killed a cock Pheasant and is feeding, while 'mantling' with spread wings over her prey in protective threat.

such as some warblers and tits, and swift-flying, common, aerial species such as swallows and martins which are normally beyond the Sparrowhawk's skill. Male and female Sparrowhawks between them can, however, take anything from a Goldcrest to a full grown Woodpigeon, the larger items overlapping the Goshawk's food range. The Goshawk's mode of hunting is essentially similar to that of the Sparrowhawk and of most other accipiters. Occasionally, however, accipiters perform surprising feats; I have seen an African Goshawk pursue a high-flying Red-eyed Dove at speed, and catch it just before it could escape into forest.

Eagles such as the Golden Eagle, Bonelli's Eagle, or Martial Eagle which habitually kill gamebirds, or crows, bustards, and ducks, normally catch them on the ground, not in flight, though perhaps they catch flying birds more often than is supposed. Red Grouse fly when an eagle appears, however, so that we assume they are safer flying than hiding in heather. Bonelli's Eagle sometimes uses cover, much like a Sparrowhawk, to approach close to possible prey and may then catch it in flight just as it takes off. This swift-flying eagle has also been seen to pass under a Jackdaw, turn upwards, and catch it in flight in the manner of a Goshawk. Martial Eagles are probably too big to catch flying birds; their special technique is long-distance attacks at great speed from a high, soaring pitch, whence they can see their prey kilometres away. I have seen a Martial Eagle catch one of a flock of feeding guineafowl from at least 5 kilometres, appearing suddenly at speed and grabbing one before they could escape.

The most spectacular and dramatic bird hunters are the big, swift falcons, Lanner, Saker, and especially the Peregrine. The Hobby and Eleonora's Falcon are even more graceful fliers but kill much smaller birds. The prey is sometimes struck dead in the air but more often the falcon seizes it in its foot and carries it to a perch. Such attacks are positively breathtaking in their speed and skill. Although we regret that the luckless prey is dead, few can withhold a feeling of admiration at the headlong rush of the stooping falcon, the split-second timing and accuracy, the smack of the impact, and the leisurely and victorious descent of the falcon on its stricken prey. To see such an attack is a red-letter moment, and most do not grudge the falcon its success. The prey may be as big as or even heavier than the falcon but more often is smaller. Most of the British Peregrine's prey is domestic pigeons or smaller birds, of two-thirds to less than a quarter of the Falcon's weight, and German and French Peregrines feed mainly on jays, thrushes, and other larger passerines.

When you see these magnificent aerial attacks casually, it often appears that they are the snatched opportunity of a moment and at least twice when I have seen a Lanner kill in flight, they took, respectively, a fruit bat and a bee-eater disturbed by me. More careful observation of about 400 attacks by the Peregrine in France by R J Monneret shows that such attacks are frequently well planned and executed from long range. A perched falcon, seeing possible prey, tightens its plumage and bobs its head up and down. If the quarry is a flying bird, it may wait till it is out of sight or behind an obstacle before starting in pursuit. Three-quarters of all attacks start from an elevated perch 200 to 300 metres above the prey, and from low cliffs the falcons must soar to gain the necessary advantage of height. The falcons killed at distances varying from 200 metres to 4.5 kilometres from the start of the attack, the kill at 200 metres being an unwary Stock Dove 'playing' on a rock. Most kills were made at 500 to 1500 metres range. The Peregrine ignores passing birds which are not vulnerable, or are too close, but seizes any good chance to kill one in a vulnerable position some distance away. In one particularly beautiful instance, a pair combined to kill a pigeon crossing a river 2.5 kilometres away. The male struck and killed the pigeon in mid-air, and it would have fallen into the water and been lost, but the female, coming 100 metres behind, caught the pigeon as it fell. About one attack in ten seen was successful.

Although a female Goshawk, weighing 1200 grams, can

occasionally kill hares four times her own weight, such mammals are normally taken only by the bigger eagles. The Golden Eagle prefers to live almost entirely on rabbits, hares, and mammals of similar size, and 99·9 per cent of the prey of the Verreaux's Eagle in Rhodesia is Rock Hyrax, weighing up to 5 kilograms. Most such big, mammal-eating eagles weigh 3500 to 5500 grams and habitually kill prey of 1000 to 5000 grams, that is, a good deal lighter than themselves.

The largest and most powerful of all is the South American Harpy, weighing 4500 to 8000 grams but no detailed records exist of the prey this huge bird can kill. Its smaller, African, ecological counterpart, the Crowned Eagle, weighing 3500 to 4000 grams is said to feed mainly on monkeys. Its staple food in Kenya is a small forest antelope, the Suni, weighing about 4000 to 4500 grams varied with smaller Tree Hyrax, some monkeys, adult duiker of up to 9000 grams (twice the Eagle's weight), and occasionally calves of still larger antelopes, the largest recorded being a young Bushbuck of at least 18 kilograms, perhaps even 25 kilograms, more than four times the Eagle's own weight.

I do not know quite how the Eagle kills such animals but having been struck by a female I can testify to what it feels like. She doubtless meant to scare rather than kill me, and she took mostly shirt but she struck me a heavy blow on the back and made three deep talon wounds. It was the force of the blow that I remember more than the piercing of the talons, which only hurt afterwards. An unsuspecting monkey or hyrax could have been knocked clean off the branch by such a blow. See page 208.

Monkeys voluntarily descend to the ground to avoid this Eagle. One luckless Sykes' Monkey was caught and killed against a chainlink fence in a neighbour's garden. Once caught, the frightful force exerted by the clutch of the thick, strong toes and the huge hind talons could clearly still the struggles of a monkey or small antelope almost at once. A Crowned Eagle can drive its hind talon through a double-thickness, horsehide falconer's glove. Perhaps, when stilling the struggles of a kicking, 20 kilogram Bushbuck calf, that hind talon can be positioned to penetrate between vertebrae of the neck till the spinal cord is pierced and the animal paralysed. When I examined the body of the large Bushbuck, I found deep talon punctures suggesting that this had occurred.

Such big animals cannot be carried entire but must be dismembered. A Golden Eagle, with weights attached to its feet, could only fly with about 1 kilogram and the maximum such an eagle can lift against a favourable wind is about 3·5 to 4·5 kilograms. I have sometimes found half a Mountain Hare lying on Scottish hillsides, clearly left by an Eagle. Such prey is normally broken in half just behind the rib-cage, the Eagle taking a good meal of flank, liver, heart, and lungs in the process. The Crowned Eagle skilfully dissects limbs of the larger kills and caches them in trees some distance away till needed. The Sykes' Monkey killed against the fence had one hind leg skilfully removed within an hour, and with a Bushbuck calf, killed 2·5 kilometres from the nest, both male and female brought a leg to the eyrie next day, though these legs were not near the kill where I found both the adults the previous evening; they had evidently been hidden in trees overnight. Some day, I shall be there and see exactly what happens!

Larger animals, weighing more than, say, 15 kilograms cannot be killed by any bird of prey but are eaten as carrion, either as the remains of a large, mammalian carnivore's kill or more often after dying a natural death. The Golden Eagle and several northern sea eagles eat much carrion in winter; 40 per cent of the Golden Eagle's winter diet may be carrion. The caracaras and Bateleur also eat carrion, but the carrion specialists are the Old and New World vultures which eat little else and, because they need not kill live animals, have lost the powerful grasping feet and talons of eagles.

Fig. 24 A Golden Eagle and a sheep to the same scale; the Eagle could not possibly kill the sheep which is six to seven times the Eagle's weight.

Carrion eating is most characteristic of the tropics where the greatest variety of carrion-eating vultures and possible food occur. It is also typical of the subtropics, however, and in the Arctic and temperate areas is important for the survival of the Golden Eagle and sea eagles in winter. In a cold climate, a stranded dead whale will evidently keep for weeks and any number of Bald Eagles can gorge there day after day. In the Rocky Mountains, a dead deer lying in the snow will also keep for weeks but may be frozen hard and normally inedible. A warm, sunny day, however, may thaw it enough for a Golden Eagle to pick off a meal. In such areas, only one or two large eagles feed on carrion and any available keeps well. In the tropics, or even the subtropics in summer, carrion becomes putrid in a few days or hours, and is usually eaten by large numbers of vultures together. Contrary to opinion, they usually eat it fresh. A carcass only lies around for even a few days when there are not enough scavengers available to clean it up. A hundred vultures or more may descend on a carcass of a zebra and consume everything but the larger bones and some skin in an hour or two. As many as six different vultures may feed at a carcass but they are specialized for different functions so that they need not compete, and all can obtain something.

Individuals among a score or more of the same species must, however, compete for precedence. An arriving Griffon Vulture lands some distance from the carcass and bounds towards others feeding there, wings spread, and long scapular feathers erected like a pair of devil's horns. Or it lowers its head and advances with legs and feet outstretched. Rüppell's Griffons in Africa fight with one another and with White-backed Vultures more than with other vultures round the same carcass. Lappet-faced Vultures threaten Griffons by a slow advance, head lowered, the specialized, lanceolate ruff feathers erected, and the tail raised vertically. The large White-headed and Lappet-faced Vultures frequently arrive at a carcass before the Griffons but then rob and threaten these later arrivals. Lappet-faced are the most aggressive African vultures and the general rule is that the aggressor wins and the defender gives way. Thus, in what appears to be a melee of feeding birds, there is an order of precedence with a few dominants actually in possession, others which have fed and are for the time being satisfied, hungry subdominants, and perhaps a new arrival or two which at once rushes in and attacks to become dominant itself until another such challenges it. Smaller, weaker vultures such as the Hooded Vulture and Egyptian Vulture will not normally attack large vultures but may attack one another.

The Lammergeier, specializing in bones, is a sort of avian hyena. It cannot eat the biggest bones like a hyena and it does not eat only bones but it undoubtedly does eat bones and drops them on rocks to break them open. It is a specialist, slow glider with very long wings and low wing loading but remarkably agile for its size and span. After collecting a bone, it carries it in its feet to a suitable area of bare rock or scree where the bone should break when dropped. The Lammergeier approaches the dropping zone downwind perhaps to increase speed, sometimes but not always dips or dives in flight, perhaps to increase accuracy of aim, and releases the bone 30 to 60 metres above the rock where it falls with an audible smack and perhaps splits into fragments. The bird can hit a 10 by 10 metre rock from 100 metres above. The Lammergeier, meantime, has quickly turned into wind and lands upwind with a quick, gentle fanning of its long wings. It must land without delay because it is often followed by more manoeuvrable crows and ravens hoping to steal scraps of bone or marrow.

If the bone splits, the Lammergeier can swallow whole pieces of bone 20 centimetres long. If not, it may try again several times, and often must glide for a kilometre or two to regain the height needed for another approach and drop. After three or four successive failures, it usually gives up and favoured rocks become covered with abandoned bits of bone and are known as ossuaries. The dropping behaviour seems compulsive with any suitable object; I have watched one in Ethiopia repeatedly drop a light hare's foot (collected from the abandoned kill of a pair of Steppe

Eagles) which fell slowly like a leaf, could never have smashed, and could have been swallowed whole anyway.

At times this huge bird can be unbelievably agile. At Dessie, I watched four adults circle up for more than 300 metres, one with a bone. Two then drifted away and finally one of the other pair dropped the bone. The second bird instantly dived vertically after it and, because of its superior weight, overtook and passed the falling bone. It seemed to flick its head to judge its chances as it passed, then turned up and neatly caught the bone after diving about 250 metres.

Many unrelated carrion-eating birds go in for piracy. Piracy must be distinguished from scavenging, which is just finding something dead and eating it. In piracy, one raptor robs another of food, whether the first raptor scavenged or killed the prey. Scavenging kites, several fish and sea eagles (of the genus *Haliaeetus*), the Tawny Eagle, Bateleur, and some vultures themselves are all pirates. In the Tawny Eagle and African Fish Eagle the impulse to piracy seems compulsive. Even if they are not hungry they may rob birds much larger than themselves but also frequently bully and rob smaller or weaker species. At carcasses where many vultures are gathered, a Lappet-faced Vulture will frequently rush at and snatch a piece of meat possessed by a Griffon but may then itself be robbed by a jackal hanging round the outskirts like a scrum-half. Piracy is usually directly connected with the carrion-eating habit, but occasionally is done by non-habitual pirates. Thus, I have seen a Verreaux's Eagle rob a Martial Eagle of its kill, and Hobbies in Holland occasionally rob Kestrels of small mammals, which they themselves would never hunt.

Having obtained a general idea of how various wild raptors kill or obtain various sorts of food it still remains to describe exactly how a kill is made. Again, in species such as the Osprey and sea eagles which hunt along watersides, the action can often be watched at close range. The Osprey normally flies 30 to 90 metres above the water and frequently poises against the wind or hovers with fanning wings. If it sees a catchable fish, it may check but often immediately dives headlong. Slow-motion films and photographs have shown, however, that as it nears the surface the Osprey throws its long legs and feet forwards beyond the head and crashes in feet first with a tremendous splash. Sometimes it stops just above the surface and swoops up again, the chance having been lost. If it plunges in, it usually catches its fish. It often takes fish in shallow water, but photographs prove that on occasion an Osprey goes right in, completely submerged all but the upheld wingtips. It may then be catching a fish almost a metre below the surface. An Osprey's feet and legs are relatively thick and strong, perhaps adapted to withstand the violent impact of this method of fishing. The fish caught, the Osprey rises from the surface, shakes itself vigorously rid of droplets, and takes the prey, carried fore and aft like a torpedo slung beneath an aircraft, to a perch. If unlucky, it is robbed of its prey en route by a piratical fish eagle or sea eagle.

The methods of sea and fish eagles are different from those of the Osprey. They do occasionally plunge right in with a splash but they do not normally hunt from high, soaring flight and when they catch a fish thus they are perhaps just taking an irresistible chance. Normally, the African Fish Eagle hunts from perches, either by what I call a 'fishing sortie', a short circling flight over the water for a few hundred metres, or more often by a 'short strike', a gliding downward flight for 25 metres or less. Fish Eagles spend hours sitting on trees waiting for the opportunity to make such a strike. When the opportunity comes, it must be seized and in a few seconds the Eagle has caught its fish and is flying back to the perch with it. So far as is known, this method is typical of other sea and fish eagles.

Although less spectacular than the crash and splash of the plunging Osprey such a catch is often beautifully dexterous. The Eagle glides down from its perch, levels out close to the water and, as it passes over the fish, throws its feet forward, and with a backward and downward snatch, lifts the fish clean out of the water behind it. Evidently, such a catch can only be made close to the surface and there may be little splash. If it must, the fish eagle too will plunge right in, soaking the whole body plumage, and disappearing in a cloud of spray.

Circumstantial stories and even photographs exist suggesting that the Osprey and sea or fish eagles will occasionally attack prey far too large for them to carry, lock their talons in it, and be dragged down to drown. All such accounts are probably competely untrue. The African Fish Eagle instantly drops a fish which cannot be lifted. At Naivasha, we tried to catch them for ringing with a floating fish, hung with nooses, and attached to a thin, nylon fishing line. They would pick up the bait, but the instant they felt the resistance of the line, dropped it, never even getting a claw entangled. I cannot, therefore, see why a fish eagle or Osprey which has inadvertently seized a fish too big for it does not just let go when it finds itself being pulled under. Yet these tales persist.

Sea and fish eagles do sometimes seize fish or birds too big to be immediately lifted clear of the water. They then float on the surface, usually with wings spread, and sometimes apparently strain, as if, beneath the water, they were convulsively gripping the prey with their powerful feet to still its struggles. Then they flap forward over the surface of the water and usually manage to rise once they gain some speed but sometimes have to flop laboriously to shore with the body and legs still submerged. A very large fish may also be trailed along the surface, partly under water. As any fisherman knows, you can draw a played-out trout to the bank on fine nylon but try to lift it out and the nylon breaks, with the greater effective weight of the fish.

Few other wild raptors give the same chances for close-up observation as fish eagles and the Osprey, and even falconers or austringers (who keep and fly short-winged hawks or accipiters) cannot usually observe precisely how their charges kill their quarry, though it often happens at much closer range than with wild birds. Austringers often have better chances to see how their hawks kill than any number of hours of observation in the wild state would provide. The hawk is always thrown from the fist when it has an advantage over the quarry, and quarry which might naturally escape is often beaten out again for the hawk to kill. Though evidently we cannot regard such kills as wholly natural they do at least give a very good idea of how an accipiter does it.

Recently, G E Goslow has used sophisticated, high-speed photographic equipment, to record the exact mechanics of the attack and strike in two accipiters, Cooper's Hawk and the American Goshawk; a buteo, the Red-tailed Hawk; and three falcons, two of them, the Prairie Falcon and the Peregrine, aerial killers of big birds, and the third a killer of small mammals on the ground, the American Kestrel or Sparrowhawk. Thus, his observations give a very clear idea of the precise method of killing adopted by some of the most successful types of raptors. By taking very careful measurements of the raptors first, and filming the attacks at 800 to 1000 frames per second with exposures of $\frac{1}{3000}$ to $\frac{1}{5000}$ second against a prepared background, he could not only estimate the exact speed and position of the raptor as a whole, but of its different limbs and how it used them.

Both accipiters attacked in essentially the same way. They first gained speed by vigorous flapping when approaching the prey, ceased flapping, the Cooper's Hawk at 3·5 to 4·5 metres and the Goshawk at 7·5 to 9 metres. Still closer, the hawks swung the body upwards to bring the pelvis beneath the head, spreading the tail to brake, and at the same time threw the feet forward hard. At the actual instant of strike, the feet were travelling towards the prey at almost or more than twice the speed of the head. In Cooper's Hawk, the relative velocities were 4·8 metres per second for the head and 11·4 for the feet; in the Goshawk 14 and 22·5 metres per second. Head, pelvis, and feet were all travelling at different speeds; in Cooper's Hawk these speeds were, respectively 4·8, 9·5, and 11·2 metres per second.

The human body is not flexible enough to simulate such movements, but if you imagine running at a wall with your fist rigidly extended and just hitting the wall, and then doing the same, but this time holding your fist back to punch the wall with all your might when within range, you can get the idea. You would certainly hurt yourself much worse in the second case. Just so, by thrusting the pelvis and feet forward, the striking accipiter delivers a violent blow at the prey and does not just grab it with outspread feet. This is why I felt such a violent blow when struck by the Crowned Eagle; she actually hit me as she passed over.

At the moment of impact, the raptor's tail and wings were thrown forwards and spread so that the direction of travel was upwards and forwards rather than simply horizontal. Even so, the momentum carried the hawk and prey some distance and if the strike was not quite straight, the hawk might spin round through almost a half-circle before coming to rest almost facing its starting point. A Cooper's Hawk attacking a pigeon from below, threw its pelvis and feet so far forward that it seemed to lie on its back and strike upwards. To kill their prey these accipiters characteristically used a kneading action of the talons, something like that of a contented cat on a lap.

The Red-tailed Hawk essentially behaved like the accipiters, but at much slower speed, as field observation would suggest. Its head, travelling at 4·25 metres per second was moving little slower than that of Cooper's Hawk but its feet at only 6·5 metres per second were moving only about half as fast as those of the Hawk. At the strike, the body axis was nearly vertical as in accipiters but the whole action was much less violent. Both accipiters and the buteo could compensate for last minute changes of direction or prey movement by small adjustments of wings and tail; they might strike and grip with one or both feet, and slew round or face the same way accordingly.

With the big falcons, the difficulties of high-speed filming were increased because they were free flying at live prey or a swung lure. The Prairie Falcon always seized the prey but the Peregrine sometimes struck the prey without gripping it in the traditional, grand manner. When seizing prey, the falcon approaches from behind, level with the quarry, and then flings pelvis, feet, and legs forwards and upwards to grip; the action resembles that of striking accipiters but the relative velocities of falcon and prey may be less. A Peregrine stooping at prey from high above passes over almost level with its prey with legs flexed, and strikes downwards with all four toes fully extended, closing them immediately afterwards. The brief contact, of a few milliseconds delivers a glancing blow with the powerful hind claw immediately knocking the prey to the ground. In the cases timed, the Peregrine was travelling at only about 18 metres per second but the prey was killed instantly by the force of the blow.

The American Kestrel descends at a steep angle with wings partly flexed. Just before impact, the wings and tail are spread to brake and the pelvis is swung forward to extend the legs below the beak. The feet are again travelling faster than the pelvis and the toes are not spread until just before impact when the bird thrusts them sharply downwards – stamping in the air, as it were. The falcon then drops on to its heels and bounces back up again, the wings being thrown forward and back after contact.

Fig. 25 Peregrine Falcon killing in flight with talons open (after G E Goslow).
Reproduced by courtesy of Auk *magazine.*

Although these experiments were carried out with captive birds under controlled conditions, there is no reason to suppose that the hawks behaved differently to wild birds. The main feature, not clear from naked-eye observation, or even slow-motion films, is that all these raptors deliver a blow in motion as well as just crashing on to their prey, by swinging the pelvis and feet forward faster than the head and body are travelling. The shock at impact to the prey must be great, quite apart from the subsequent vice-like grip of the sharp talons. The relative shock depends on the weight of the bird and the speed at which it or its feet are travelling. An American Goshawk, 10 to 20 per cent lighter than a Red-tailed Hawk, nevertheless delivers a much greater shock because it is travelling almost three times as fast (14 metres per second as opposed to 4·25) and its feet nearly four times as fast (22·5 rather than 6·5 metres per second). The energy delivered varies as the square of the velocity, so that a goshawk would strike prey about ten times as hard as a buteo of about the same weight. This explains why goshawks kill relatively much heavier prey than do buteos.

The pictures apparently also settle a long-standing argument about Peregrines. When killing, the Peregrine strikes with open feet, and then closes them. It does not, as has often been supposed, deliver a blow with its foot clenched. A Peregrine may repeatedly attack and strike at the head and neck, or it may inflict a severe wound at the first contact. The weight of the Falcon is not directly above the quarry at the moment of strike, and the blow is a glancing one delivered with the hind claw, otherwise the Falcon might break its own strong leg. Again, the central feature is the blow, not the speed of the attack.

These experiments show quite clearly how all accipiters and their near allies, all buteos, and several small eagles that behave the same way, and most of the commoner falcons kill their prey. Methods will obviously vary and wild birds probably perform better than captives, especially when experienced. Further, high-speed, experimental photography with snake eagles and others which kill specialized prey, now that the technique is known, would be valuable. I would expect the action of a snake eagle to resemble that of a kestrel rather than a buteo, delivering a hard, downward blow at contact.

Having killed, a falcon breaks the neck of its quarry, but accipiters and their allies do not. A chanting goshawk (essentially like an accipiter), which seized a full-grown guineafowl within 20 metres of me one morning, first dragged its quarry protesting out of the bush where it had hidden, and then tore at the neck till it had killed the bird. It fed only off the neck; I ate the rest. In this case, the bird killed was at least three times the weight of the killer but with smaller birds the head would be torn off and swallowed before the body was carried away.

Where a hawk or eagle kills, is found the plucked feathers or fur and usually the stomach of large animals. If the stomach is packed with vegetable matter, it is rejected. The raptor then eats into the body cavity, consuming intestines, liver, heart, and other organs before eating much solid flesh. Fish eagles first break into the soft belly of a fish. Small prey may be swallowed whole at once; for instance, a vole caught by a Common Buzzard. If it is very large, prey may be dismembered and carried away in bits, or it may lie where it is for several days, the killer returning to it at intervals. The Crowned Eagles that killed the 20 kilogram Bushbuck lost most of it, eaten that night by a mammalian predator, perhaps a Leopard. In the same forest, a young Martial Eagle returned to its Suni kill for four successive days until only the head and the long limb-bones remained uneaten.

The amount eaten depends on appetite. Raptors do not normally kill unless they are hungry, but they sometimes will, and in that case may not eat at once. All are capable of eating much more than their required daily ration at one gorge, especially such birds as vultures with their life of alternate feast and famine. A Golden Eagle can gorge about 1500 grams at once, about six times its normal daily need. The bigger the raptor the more it can normally eat at one meal and the less often it actually must feed. Smaller raptors require more food in relation to their body weight per day than do large ones, and they need more regular meals.

One important question remains – how successful are hawks, eagles, and falcons in their hunting? Opinions vary on this because few people have ever watched hawks systematically for long enough to obtain enough data, and those who have disagree on whether an attack is for real or not. Gustav Rudebeck, in his pioneer Swedish studies concluded that Peregrines succeeded in killing only once in fifteen attacks (6·675 per cent) and Merlins only once in twenty (5 per cent). In 469 Osprey attacks observed by

Lambert in Nova Scotia, however, 419 (90 per cent) succeeded. Again, most others who have watched Ospreys conclude that they are extremely successful, though my own observations on the Kenya Coast suggest that only one attack out of three or four succeeds. Everyone I know who studies large eagles or large falcons concludes that they can kill almost at will when a good chance occurs. Were this not so, those who have spent months on end walking thousands of kilometres in the Highlands of Scotland would surely have seen Golden Eagles kill more often; and the Verreaux's Eagles of the Matopos would have been seen to kill hyrax more often.

A check in flight is not necessarily an attack. An Osprey which checks and hovers is just assessing its chances. If it drops a metre or so and then gives up, that is hardly an attack either but if it goes right down to the water that is clearly an unsuccessful attempt to catch something. Mock attacks are also common, especially among big falcons and eagles. In Glen Callater I once saw a Golden Eagle stoop magnificently at a Herring Gull, and miss it by a fraction, terrifying the Gull. The Eagle could clearly have killed the Gull if it wished. It often seems that the Peregrine Falcon just cannot resist having a mock swipe at a passing pigeon, as if to scare the daylight out of it!

Most systematic observations show, however, that most raptors make several unsuccessful attacks for every successful one. Dartmoor Buzzards vary in success, both from day to day and between neighbouring pairs on the same day. In one pair, the male hunted for eleven-and-a-half out of sixteen hours, the female for six, but the time spent hunting for the daily food need depended partly on whether they caught a rabbit or not. The Craigheads in winter in Michigan concluded that buteos spent most of the day actually hunting but at Lake Naivasha my observations show that fish eagles spend from 0·2 to 8 per cent of daylight actually flying to hunt, averaging about 1·3 per cent. Thus, at one extreme there is the impression of a bird struggling hard to get enough to eat, and at the other, of a lazy creature that can eat without difficulty more or less at will. Even in the short time they spend hunting, fish eagles make about five unsuccessful attacks or strikes for every successful catch. Piracy is more successful – they seldom fail, which no doubt partly explains its compulsive nature.

More data, systematically gathered from long-sustained observations are needed to elucidate this subject. Species living in temperate or cold climates, however, may have greater difficulty in obtaining their food than tropical raptors. This, apart from the effects of sheer cold, could be another reason why raptors are so much more abundant and varied in tropical climates than in the frozen north.

Migration
and nomadism

Both these activities are largely connected with fluctuations in food supply, sometimes but not always caused by changes in temperature. Migration is defined as a regular movement carried out by the whole, or a part of a bird's population from its normal breeding range to its wintering or non-breeding range. Most migrants were first noticed in temperate climates, so that this non-breeding area may be called the winter range, even when discussing entirely tropical migrants. A migration involves a yearly journey of at least 400 kilometres each way, usually much further. This can include everything from the long-distance movements of Swainson's Hawk from the Great Plains to the Argentine pampas and back to the short but still regular movements of Grasshopper Buzzard-eagles in tropical west African savannas.

Nomadism, on the other hand, is essentially irregular movement in varying numbers, in almost any direction, perhaps in response to fluctuating food supply, but often unexplained. A human nomad is one who wanders about with his flocks in deserts and steppe; and this is close to the truth among raptors too. While migration may occur alike in temperate, subtropical, or even tropical climates, nomadism is more characteristic of arid and semi-arid, subtropical and tropical areas. Neither migration nor nomadism is typical of the very stable, warm, tropical forest habitat but in savannas and plains both may occur even in the tropics.

There is really no absolute division between migration and nomadism, and basically migratory species may also be partly nomadic. The Rough-legged Buzzard, completely migratory from tundra and taiga breeding range, winters mainly in open parts of the temperate zone and here it is nomadic, moving from place to place, probably according to food supply, often semigregarious at roosting places or on good hunting grounds. Territorial in its breeding quarters, it is much more sociable during its nomadic winter life.

Migration is comparatively easy to observe because it is conspicuous and regular; you come to expect it and watch for arrivals and departures. Nomadism is much less known, understood, and recorded because it is irregular and usually unpredictable. A rat plague may attract more Black-shouldered Kites than usual to an area but by the time it is noticed the plague may have died down and the Kites nomadically moved elsewhere. Time may show that many more species are nomadic than can now be recognized as such. Moreover, notably in south America and Australia, bird movements are so poorly recorded that we often cannot say whether a species is nomadic or migratory or both. We may only say that it is not a permanent resident.

Even species usually considered permanent residents, such as the African Fish Eagle, also wander. When Lake Paradise on Marsabit Mountain (160 kilometres from the nearest water at Lake Rudolf) filled up, Fish Eagles occupied it; it contains no fish but they eat water birds – also wanderers. A forest-loving Black Sparrowhawk has been caught aboard ship off west Africa. Such birds are neither migrants nor even nomads, just wanderers. There is no certainty that any species, however unlikely, may not occasionally occur outside its normal haunts.

Most birds of prey are resident. The proportion of migrants, complete or partial, decreases steadily from the Arctic to the tropics, and migration does not occur inside tropical forests. Of the 287 raptor species, only twenty-three are completely migratory throughout their range. Another twelve are mainly migratory and thirty-nine partly migratory. About seventeen species can be regarded as definitely nomadic but New and Old World vultures also are certainly locally nomadic, while the scavenging caracaras and milvagos are doubtful cases with their obscure movements. Just over a quarter of all raptors migrate at all and only a twelfth are wholly migratory.

I have listed wholly and partly migratory and nomadic species in Appendix IV. Wholly migratory species include two insectivorous American kites and the Black Kite, three harriers, two accipiters, two grasshopper buzzards *(Butastur)*, three buteos, four booted eagles, and five falcons. The mainly migratory species include the Osprey, the Honey Buzzard, the American Swallow-tailed Kite, the Plumbeous Kite, Red Kite, one harrier, one accipiter, one buteo, one eagle, and three falcons. Partially migratory species include the Turkey Vulture, two cuckoo falcons, four sea eagles, three Old World vultures, one snake eagle, three harriers, one chanting goshawk, five accipiters, ten buteos, three eagles, and six falcons. The groups with the highest proportions of migrants are the largely insectivorous kites and honey buzzards, seven of about fifteen species; buteos with thirteen of twenty-six species, and falcons with fifteen out of thirty-seven species.

The twenty-three completely migratory species include one Arctic, two temperate, fifteen temperate to subtropical, and five exclusively tropical species, with two species, the Mississippi and Black Kites, apparently migratory throughout their temperate and tropical range. All tropical migrants are savanna species except the Sooty Falcon which breeds on Red Sea islands and winters in Madagascar. Although most raptors apparently migrate to escape the effects of

Far left A Martial Eagle on its kill; its excitement is self-evident. Though normally feeding on gamebirds, it has here killed something more substantial.

Left A Tawny Eagle intently watching a rodent hole at a road edge; although gorged, the bird has been unable to resist making an attack when a good chance came.

Below Hawks are seldom seen drinking; this Grey Kestrel of African savannas is drinking from a muddy pool; its powerful beak suggests reptile eating, though its prey is unknown.

■ breeding area

⬚ wintering areas
(immatures in Florida)

Fig. 26 Swainson's Hawk: the most complete of any American raptor migrations; all but a few immatures winter in the pampas of Argentina after breeding in North American western prairie and steppe.

cold, probably through shortage of food, in tropical species this cannot be so and it is not clear whether movements and food supply are related.

All twelve mainly migratory species live mainly in temperate or subtropical regions with subtropical or tropical races. Three are basically temperate and eight temperate to subtropical, while the Plumbeous Kite is mainly tropical. Of the thirty-nine partly migratory species, seven are Arctic to temperate, none wholly temperate, eleven temperate extending to the subtropics and sometimes the tropics, seven basically subtropical, eight subtropical to tropical, and four wholly tropical. Most of the partial migrants extend through two or more main vegetation zones, having a northern population (sometimes also a southern) which migrates and a resident subtropical or tropical population. The four partial tropical migrants are all exclusively tropical including one rather doubtful species, the Ovampo Sparrowhawk.

The completely migrant species are the most specialized and perform the most remarkable movements. The Black Kite seems to be migratory or strongly nomadic wherever it occurs, even in Australia. Some kites may be present year-round in tropical areas but they may include members of several different races or populations. Thus, in Kampala, Uganda or Nairobi, Kenya in winter (October to March) migrant European Black Kites, *Milvus migrans migrans*, mingle with migrant and breeding tropical African Black Kites, *M.m. parasitus*, while possibly some unrecognizable migrant tropical African Kites from southern Africa arrive during the southern winter during April to September (when they are absent from South Africa). Its omnivorous diet, readiness to migrate or be nomadic as need be, live alongside man in numbers, and its grace and skill in flight make this, probably the world's most successful and perhaps most numerous raptor. Nor is it as unattractive as the variety of epithets applied to it suggest.

The migrations of Swainson's Hawk are probably the most spectacular and complete of any American raptor. Breeding over a vast area of plains from California and Texas to Alaska, the whole population pours through Central America in autumn, and again northwards in spring, often in huge flocks. All winter in the pampas of Argentina, an area resembling their breeding range, except a small population of immatures which have lately begun wintering in Florida. The only comparable migration among Old World buteos is that of the Steppe Buzzard, *Buteo b. vulpinus*, a northern race of the Common Buzzard, breeding from near the Black Sea north to Finland and east to Siberia, but wintering mainly in South Africa. There is no such narrow land corridor in Africa as there is in Central America so that the migration is more diffuse.

The complete migrations of the Pale, Montagu's, and

Pied Harriers are largely straight north-south migrations, resembling those of the European and north Asian races of the Marsh Harrier. Most migrant harriers do not go far south of the equator. These movements have curious features; for instance, although large numbers of Montagu's Harriers cross the Straits of Gibraltar they seem less common in west African savannas than Pale Harriers, which do not. More complete records may in time explain such questions. Two of the three completely migrant accipiters are eastern, and one middle eastern. One is a typical bird eater which has no obvious reason for complete migration but the Levant Sparrowhawk and Grey Frog Hawk are reptile or frog eaters and their food supply disappears entirely in winter. With these two we must link the Shikra, a partial migrant wholly within the tropics, which is also a lizard eater.

Migrant eagles present curious problems. The Lesser Spotted and Greater Spotted Eagles breed in rather similar habitat in central Europe and western Asia and both winter in Africa. Numbers pass through Suez and the Bosporus in autumn even today when they have become scarcer. They winter in quite different parts of Africa, however, the Lesser Spotted apparently travelling straight on beyond the equator to the southern woodlands of Rhodesia and the Transvaal. Within Africa, good records of it are extremely scarce, and of odd birds, not of flocks. At present, it seems likely that this big eagle travels from Suez to southern Africa almost unnoticed, perhaps non-stop. The Greater Spotted

Eleonora's Falcon, pale phase; this extraordinary migrant Falcon breeds on Mediterranean islands at the time of southward migration of passerines out of Europe and winters in Madagascar.

Right The hawk ringing station at Hawk Mountain in autumn; the open area is set with mist nets, and hawks are trapped, ringed, and released after obtaining data.

Below Examining a trapped male Dark Chanting Goshawk for moult; his right wing, showing emargination of the four outer primaries even in a short-winged hawk, is feather perfect.

Below Eleonora's Falcon, dark phase. One of the strangest of all migrant falcons.

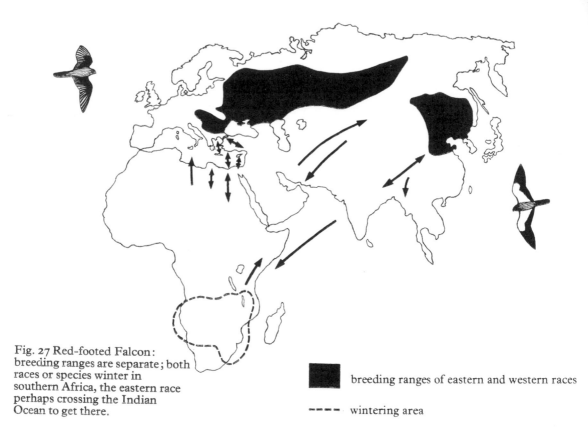

Fig. 27 Red-footed Falcon: breeding ranges are separate; both races or species winter in southern Africa, the eastern race perhaps crossing the Indian Ocean to get there.

■ breeding ranges of eastern and western races

- - - - wintering area

Eagle was said to winter mainly in north-east Africa and Ethiopia but recent good records are lacking. The commonest winter migrant eagle in Africa is the Steppe Eagle but because it does not come through Istanbul or Suez in numbers, how it reaches Africa remains mysterious.

Falcons perform the most remarkable complete migrations. Eleonora's and Sooty Falcons both breed in August, feeding on the rush of migrants southward bound to Africa, in the Mediterranean and the Red Sea respectively. In the breeding season, they eat birds but choose to spend the winter in Madagascar, mainly eating insects. We might well ask, why Madagascar? To reach Madagascar the Eleonora's Falcons from Mogador, off the Moroccan Coast, must first fly north, then east, and finally south. Although insects would be abundant inland their route is probably mainly coastal though little known. Sooty Falcons are more often noticed en route than are Eleonora's.

The Lesser Kestrel, the Hobby, and the Red-footed Falcon breed in temperate or subtropical Europe and Asia and winter in southern African woodlands. Some Hobbies winter in north India but not in eastern tropical islands. There is little mystery about the western Red-footed Falcon, the Hobby, or the Lesser Kestrel, all of which are seen on migration. The real mystery is the eastern Red-footed Falcon, perhaps a separate species. They breed in old corvids' nests in east Siberia and China, and on their south-ward migration they seem to reach Assam in numbers, and then almost completely disappear, reappearing commonly only in southern Tanzania and Zambia. It seems that they must cross the Indian Ocean direct from Assam to Tanzania because they are not recorded in numbers leaving the southern tip of India or Sri Lanka to make the crossing. Odd individuals have been recorded near Cape Guardafui, and a few pass through Kenya each year. Their southward route remains the greatest enigma of raptor migration. Going north, they are more easily seen, some passing through Afghanistan in company with Lesser Kestrels, just as the western race does in Libya.

The tropical migrants are of unusual interest because they migrate from one hot place to another. Two, the African Swallow-tailed Kite and the Grasshopper Buzzard-eagle are mainly insectivorous. They breed in the rains (May to July) in the northern, climatically drier part of their range and migrate into southern savanna in the dry season from October to March. Insects evidently provide plenty of food for them in the southern savannas; indeed, grass fires at that season help the Grasshopper Buzzard-eagle to gorge with ease. This food might decrease after the fires so that breeding may be more secure in the rains in the north.

The African Red-tailed Buzzard reverses this pattern. It breeds in the southern part of its range in the dry season from November to March and migrates northwards as soon as rains fall. Its 'winter' range is obscure. The partially migrant African Shikra, and some tropical African kites also migrate south in October, breed, and return northwards with the April rains. Thus, a rat-eating, a lizard-eating, and an insectivorous to omnivorous scavenger all behave the same way and in these cases, the movements are hard to connect with available food.

Wahlberg's Eagle is apparently a transequatorial migrant, breeding in southern savannas from Kenya to the Transvaal, from September to January, which is mainly wet in Kenya but partly dry in Rhodesia. This eagle is almost or wholly absent from southern Africa between March and August and may then be somewhere in the little-known wastes of the Sudan. A pair has recently bred at 11° north in the compound of the British Embassy at Addis Ababa – again laying in September; and a scatter of recent records now suggests that some also breed in the northern tropics. Several other birds, such as Abdim's Stork and the Pennant-winged Nightjar also migrate to and fro across the equator within the tropics.

South America lacks wholly tropical migrants but the American Swallow-tailed Kite and the Plumbeous Kite are mainly and possibly wholly migratory. The American Swallow-tailed Kite migrates northwards from the south-

migration routes → (partly conjectural)

▭ wintering areas

Fig. 28 Eleonora's Falcon: from Canary Islands and Mogador off Morocco first goes north, then east with others, and down African coast to Madagascar; routes are partly conjectural.

northern and southern
—— limits of breeding range
---- northern and southern limits of wintering range

Fig. 29 Plumbeous Kite: an intra-tropical migrant, breeding both north and south of the main range and migrating towards the centre.

Right A small clutch of eggs; the two eggs of a Golden Eagle weigh about 5 to 6 per cent of the female's bodyweight and require little energy to produce.

Below A normal clutch of six Kestrel's eggs weighs 120 grams, almost 50 per cent of the female's bodyweight and evidently requires much more energy to produce.

Far right A Buzzard at the nest with an above-average clutch of three eggs; white marked with brown is typical of most birds of prey eggs.

ern edge of its range and southwards from Florida and Mexico but it is not certain whether there is a tropical, permanently resident population. The northern and southern populations of the Plumbeous Kite migrate but it is also a permanent breeding resident in Surinam, nesting in March and April, whereas in Misiones, Argentina, migrants breed from October to November. These two kites may have a migration pattern as complicated as that of the Old World Black Kite but are more difficult to study because no recognizable races are known.

Specialized food supply is clearly the reason for migration in several mainly or partially migrant species. Honey Buzzards breed mainly in the north-temperate woodlands and here are wholly migratory, arriving late in May, breeding, and departing early in late August to September. Their total breeding season is shorter than that of a similar-sized buteo, presumably because only in high summer can they find enough wasps. In India and tropical Asia permanently resident races or related species occur. Likewise, the European Snake Eagle, *Circaetus gallicus gallicus*, is completely migratory from its subtropical Eurasian range to tropical Africa and India in winter but has two permanently resident races, Beaudouin's Snake Eagle *C.g. beaudouini* and the Black-breasted Snake Eagle, *C.g. pectoralis*, in tropical and southern Africa. These were formerly regarded as separate species but this is incorrect.

Partial migrants vary from species ranging widely from the Arctic to the subtropics, in which the entire northern population migrates to warmer climates to mainly tropical or subtropical species in which only a few of the population migrate. The former would include, for instance, the Sharp-shinned Hawk, Common Buzzard (one race of which performs a transequatorial migration), Golden Eagle, and Common Kestrel. The latter would include all four sea eagles, three European vultures, the Dark Chanting Goshawk, several subtropical buteos, and for instance, the Australian Little and American Aplomado Falcons. Between, there are species in which large numbers migrate, such as Cooper's Hawk and the Red-shouldered Hawk, and largely resident species with some migrants, such as the Hen Harrier in Europe and Red-tailed Hawk in America.

Size and weight are evidently important in migrant species. Broadly, the bigger the bird the less its need to migrate. Among the twenty-three migrants, ten (43 per cent) are small, weighing less than 300 grams, five weigh 300 to 600 grams, six weigh 600 to 1000 grams, and only two, the spotted eagles, are large, more than 1000 grams. The twelve mainly migratory species include only three (25 per cent) small species, and half are large or very large (the Tawny Eagle/Steppe Eagle group weighing more than 2000 grams). Among the thirty-nine partial migrants, ten are large (one being the Arctic Gyrfalcon) and eight are very large; twenty-four out of thirty-nine or two-thirds of all species weigh more than 600 grams. The nine small, partial migrants are mainly insectivorous kites or small falcons; one (the Little Falcon) is hardly migrant at all and four migrate only within the tropics. Thus, although there is a general correlation between small size and migratory habits, there are interesting exceptions, and falcons, regardless of size, seem more inclined to migrate than any other group.

Very large partially migratory sea eagles of the Arctic or cold climates may easily obtain winter food on seashores. Some of the Florida population, breeding in Florida, migrate north, and some of the northern population of American Bald Eagles migrate south. A Bald Eagle seen in central North America in summer may have come from Florida, or be a resident, but not an Alaskan migrant. Northern and southern populations of some other species migrate from the extremities towards the centre of their range but the Bald Eagle is unique in migrating entirely within the northern hemisphere. Northward migration in the southern hemisphere is generally slight, because the really cold southern land areas are relatively small.

It is also interesting to compare what happens on the various continents. There is a fair amount of migration from temperate to subtropical North America and some from the subtropics to the tropics but only a few raptors migrate beyond the barrier of tropical forests. Knowledge of such movements in South America, however, is very incomplete. In Europe and western Asia, with the huge tropical land-mass of Africa available southwards, migration, at least to the northern tropics and equator, is common and at least four species migrate mainly into the southern tropics and subtropics. In eastern Asia and China, however, migrants may move west and south into Africa rather than winter in India and Burma, though some do. Asian Lesser Kestrels and Red-footed Falcons mainly winter in Africa. The European and west Asian Steppe Buzzard, *B.b. vulpinus*, migrates south to the Cape of Good Hope but the Japanese Buzzard, *B.b. japonicus*, breeding in east Siberia and north China, winters in continental Asia and does not reach the East Indies in numbers. The Japanese Lesser Sparrowhawk and the Grey Frog Hawk, however, both cross large stretches of water to reach East Indian islands formerly part of Asia. Having done this, they apparently balk at the comparatively narrow sea crossing from Celebes to New Guinea. In other words, they do not normally cross Wallace's Line, which divides the Australasian from the oriental faunas.

In Australia very little true migration is observed. Tasmanian Marsh Harriers migrate northwards across the Bass Strait in winter and some Spotted Harriers and Little Falcons also migrate. Most other movements are nomadic.

Right An Osprey landing at its nest with small young; Ospreys and other large aquatic species tend to build huge nests used year after year.

Right Small chicks cannot feed themselves, but must be fed; they are attracted to reach for anything red. Here a female Shikra feeds her young.

Far right As the chicks grow the female Sparrowhawk remains on guard; she and the chicks are still wholly dependent on the male as a provider.

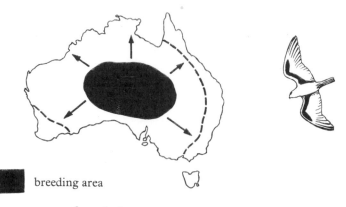

Fig. 30 Letter-winged Kite: a nomadic species, breeding sporadically in the arid centre of Australia and moving irregularly to other areas.

■ breeding area

--- extent of wanderings

I have listed seventeen species as definitely nomadic, some of them also truly migratory in parts of their range. The most typical nomads are the elanine kites, some harriers, and some of the smaller falcons in the steppes of Asia and Europe. These mainly feed on fluctuating rodent populations supplemented by insects. Some species normally resident in pairs in their breeding range are more gregarious and nomadic in winter. Thus, in winter, Common and Lesser Kestrels are found together in flocks in African savannas and several buteos tend to nomadism in winter quarters. The Bateleur and the Secretary Bird, inhabitants of tropical savannas, are probably nomadic because their numbers certainly fluctuate, in the Bateleur perhaps in response to local weather changes. Tropical African Tawny Eagles are certainly nomadic, while the northern races (Steppe Eagles) are strongly migratory from their breeding range but, nomadic, following termites, locusts, or rats, in winter quarters.

The essential features of nomadism is unpredictability, though there are a few useful pointers such as greater sociability in nomadic species. They tend to be gregarious or semigregarious, either at roosts or in breeding areas, or at least not so strongly territorial. A species may differ in behaviour in different parts of its range; for instance, some Kestrels are solitary breeders but in eastern Asia become colonial and nomadic. From what little is known, the Australian Letter-winged Kite seems highly nomadic, breeding sporadically in colonies associated with rodent outbreaks. Black-shouldered Kites also tend to roost communally, and Hen Harriers, Montagu's Harriers, and Pallid Harriers all roost communally in winter, either in temperate or tropical climates, and some show unusual territorial behaviour, notably polygamy, in summer.

Nomadic habits perhaps have evolved to help populations of certain raptors habitually dependent on rodents, grasshoppers, or both, to roost and breed close together without territorial conflict when prey is locally abundant. Much more remains to be learned about nomadism, which is generally associated with rather hot, dry places where such rodent and locust outbreaks are a regular feature, and also to some extent Arctic regions and species. Kestrels, harriers, and especially the elanine kites seem to be good species for study to help elucidate these problems.

Though nomadism is unpredictable, large-scale migration is normally both predictable and sometimes easily observed especially at points of land just before a short sea crossing. Raptors, especially soaring species such as buteos and eagles, will not normally cross large bodies of water, which provide no thermals and force raptors to use the flapping flight to which they are anatomically ill-fitted. Thus, they tend to pile up in the approaches to such crossings and cross only when conditions are right. One day, thousands cross, the next, only a few.

In Europe, Falsterbo at the tip of Scania, Gibraltar, and the Bosporus are famous. Most visible migrants enter and leave Africa through Suez or Gibraltar. Though there is no such sea crossing between continental land masses in America, movements along the Central American isthmus are not so well documented. In eastern Asia migrations usually follow the coast and there is no appreciable raptor movement to Australia. The best places to see raptor migrations are in Europe and in Africa at Suez or Cape Spartel opposite Gibraltar.

Inland, soaring raptors prefer to follow escarpments which provide lift and they fly round big lakes such as the American Great Lakes or Lake Victoria in Africa as if they were seas. Hawk Mountain in Pennsylvania is a classic example of a ridge situation, while the ringing station at Cedar Grove on the west shore of Lake Michigan has produced remarkable results in recent years. In Africa the trough of the Great Rift Valley is undoubtedly important, but little studied. In featureless country, migration routes tend to be ill-defined but any isolated mountain top is likely to be used by soaring species to gain lift preparatory to a long, high-speed glide.

Several favoured places now have permanent bird observatories, manned by professionals aided by enthusiastic amateur observers. Nowadays, visible migration can sometimes be correlated with invisible migration recorded by radar which can produce startling results. Eider Ducks, for instance, have recently been detected crossing the southern part of Scandinavia at great height and it may seem less surprising that a powerful soaring species such as the Steppe Eagle reaches Africa virtually unobserved. This is actually unexpected because most soaring species are seen chiefly at sea crossings. Certain species, such as harriers, which are adapted for flapping flight, may be readier than others to make wide sea crossings, and recent observations at Malta have shown that Honey Buzzards also use sustained, flapping flight. Even harriers at Cape Spartel prefer to wait till a favourable wind allows them to cross without flapping.

All known regular raptor migration takes place by day so that it is easy to watch. Moreover, experts can tell from weather forecasts when a good passage is likely and large numbers of watchers then converge on Falsterbo to see the movements. Counts of different species passing from day to day and year to year can be correlated with the effects of weather on migration. Migrants can be trapped, weighed, measured, sexed, their moult noted, and released again after ringing. The best observations, naturally, come from well-manned stations in north Europe and North America but very good observations have also been made at Gibraltar, the Bosporus, and Suez. The straits of Hormuz and Bab-el-Mandeb are obvious places to watch in future.

Birdwatchers at Falsterbo watching soaring hawks; 100 per cent enthusiasm for the spectacle.

A detailed discussion of such results would take up the rest of this book, and I can mention only a few points of general interest. The numbers seen can never be a true estimation of the total population because . large numbers certainly pass unseen – suspected earlier and now confirmed by radar. Moreover, the numbers of one species passing from year to year can vary greatly and may not agree with estimated breeding populations. Counts give a general idea whether a population is increasing, decreasing, or static, however. At the Bosporus, the numbers of Greater and Lesser Spotted Eagles have apparently decreased drastically since the end of World War II. At Hawk Mountain, the disastrous effects of pesticides on certain American raptors have been confirmed by decreasing records of Peregrines and Bald Eagles. Such results need rather cautious interpretation but general trends can certainly be noted, giving early warnings of a threat to a species. For instance, the

Merlin is evidently declining over much of Europe.

Some raptors feed on migration, others apparently travel fasting. Broadly, accipiters, harriers, and falcons feed on the way, while large, soaring Honey Buzzards, buteos, and eagles either fast or feed infrequently. Species which feed en route are likely to give good chances near migration stations to observe hunting success. The concentration of small birds and raptors alike could lead to unusual behaviour. Work at Cedar Grove has shown that raptors will attack Starling and pigeon baits whether hungry or not, so that at a place like Falsterbo, many apparent attacks are likely to be false. This could, for instance, help to explain the very low rate of success observed by Rudebeck in Sweden.

There are curious anomalies and exceptions to any such general rules. We would expect small, insectivorous, eastern Red-footed Falcons requiring to feed often, to be visible in numbers for that reason. Hobbies and Lesser Kestrels which perform similar, if slightly less extensive, migrations from northern Europe to southern Africa are seen feeding all along the route. Yet eastern Red-footed Falcons disappear! It is less surprising that a large, soaring species such as the Lesser Spotted Eagle should manage to reach the Transvaal from Suez virtually unobserved because it probably needs to feed less. The Steppe Eagle, however, apparently comes from Europe to Africa more or less unobserved, whereas the Lesser Spotted is easily observed leaving Europe. These cases all suggest exceptions to the general rule of diurnal migration because surely Red-footed Falcons must take several days and nights to cross the Indian Ocean from Assam to Tanzania (if they do), and even Lesser Spotted Eagles would take several days to travel from Suez to Rhodesia in continuous flight.

Low wing loading may be correlated with long migration. Honey Buzzards, which may have lower wing loading than buteos of comparable weight are the largest species regularly to cross the Mediterranean via Malta but they partly use flapping flight. Fasting could reduce wing loading, both by actual loss of weight and the lack of crop contents. A large, soaring raptor might thus reduce its wing loading by 10 to 20 per cent. This should assist earlier use of smaller thermals and a longer flying day, more rapid climbing to height on any available current, and a gentler angle of glide once the raptor leaves the thermal to travel straight. Thus, it would benefit soaring species not to eat, while those that travel mainly or partly by flapping would need more food.

Honey Buzzards must compromise because they cross the Sahara and cannot stop on their journey in spring to gather wasps – there aren't any. Perhaps, like warblers, they store up fat before they leave Africa. Other aerial insectivores, such as the American Swallow-tailed and Plumbeous Kites, can probably feed aerially on whatever sort of vegetation they fly over; they do not cross any extensive deserts.

Birds of prey are not among the furthest travelled or the fastest migrants. Nor do they perform the feats of navigation or speed of waders, geese, or even Greenland Wheatears. The longest migrations are those of Steppe Buzzards from west Siberia to the Cape of Good Hope, of Swainson's Hawk from Alaska to Argentina, and of the eastern Red-footed Falcon and Lesser Kestrel to southern Africa. All these, allowing for the circuitous routes they apparently follow, are in the order of 13 000 to 16 000 kilometres each way. Most migrations are much shorter and some individuals of the same species (the Common Buzzard) migrate 16 000 kilometres and others (in Britain) remain year-round within 1 or 2 square kilometres.

In long-distance migrants, such as Steppe Buzzards, the northernmost migrants leave first, and overtake more sedentary southern birds and travel further, as ringing results prove. Steppe Buzzards from Finland go no further than the subtropics but those that follow the Bosporus and Suez route into Africa go much further. If such a bird leaves its northern range in late August, and arrives in Cape Province in mid-October, it travels 13 000 to 16 000 kilometres in six to eight weeks, at an average speed of 240 to 400 kilometres a day. Buzzards are soaring birds and throughout the route daylight will not exceed twelve hours so that they can probably only start at 8 a.m. to 9 a.m. and must alight again by 5 p.m. They would have to average only 27 to 48 kilometres per hour to do it. As they normally travel by rising on thermals to 300 metres or higher, then gliding to the next available upcurrent, such an average speed seems well within their powers. To maintain such a speed for two whole months without any food would be scarcely feasible and they must feed sometimes en route. They must average faster on flying days; this would not be difficult.

Ringing results have shed very little light on the speed of raptor migration but have clearly outlined the paths followed. For instance, Swedish Ospreys and the small British population fly down the west coast of Europe and Africa about as far as Senegal. The western population of Steppe Buzzards from Finland and other Baltic countries moves south-west through France and apparently does not cross into Africa, while the eastern population travels south-east and does enter Africa, going south to the Cape. The western Red-footed Falcon follows a more westerly course on its northward spring migration than in the autumn migration and so probably does the eastern race, unobserved in autumn but seen travelling north through Afghanistan with Lesser Kestrels in spring. In most of the very large species enough young cannot be ringed, or adults caught to obtain enough returns, but recoveries have proved that young Florida Bald Eagles migrated north to Winnipeg and Nova Scotia.

Ringing results, particularly of birds ringed in the nest, also provide the best available data on age structures of the population. With enough results, a life table can be worked out to give an idea of maximum potential age. In the European Buzzard, Common Kestrel, and others for which thousands of results are available, all such tables show that heavy mortality occurs before sexual maturity; and generally the bigger the raptor and the longer its maturity is delayed, the heavier the mortality. About 65 per cent of British Kestrels die before they breed in their second year and results show that this proportion has not basically changed for many years, though it fluctuates year by year, perhaps according to good and bad vole years.

About four out of five buteos die before sexual maturity and about two out of three Ospreys. Most mortality occurs in the first winter, and once a bird is adult, it may live several years thereafter. Until recently, the oldest recovered birds were raptors, a Buzzard of twenty-three and a Red Kite of twenty-six. These have now been eclipsed, however, by gulls and waders of over thirty.

Thus, about three out of every four large raptors reared will fail to breed and this is still the best guide we have to survival. Most recoveries of ringed birds are of individuals that have died an unnatural death, being shot, killed by cars, electrocuted, and so on, so that the method is suspect. Thus, we can calculate that if two out of three of all Swedish Ospreys do die before breeding (as the ringing results tell us) the species could not have increased as it has because those that survived could not rear enough young. Such results must, therefore, be interpreted with caution.

Similarly, records often show a disproportion of the sexes, or of adults and immatures. For instance, Goshawks moving south through Cedar Grove are mainly young birds, with more males than females. Most of the Steppe Eagles reaching Rhodesia and Botswana in Africa are immatures. Such facts, related to more detailed observations in breeding quarters, support observations that the young disperse from established territories while adults may remain in residence. Only a very large number of results could prove that there are more adult males than females in any population as a whole.

Proper analysis of such facts needs a book to itself. Meantime, to me, it is sometimes extraordinary to think that a raptor need migrate at all, and I puzzle over possible reasons why. Ospreys, which I assume come from north Europe or Asia spend August to April on the Kenya Coast. They have plenty of easily caught fish, an idyllic warm climate, and good big mangrove trees to breed in. Yet they leave, I assume to breed on some cold Siberian lake or river. I cannot think why, particularly because in Australia in similar latitudes they would not.

Breeding biology

More is known about breeding biology than any other aspect of the lives of birds of prey, for the simple reason that the need to return to the nest to incubate eggs or tend young ties an otherwise free-ranging, fast-flying, elusive bird to one place where it can regularly be located and observed. Much of what is known about several other subjects, such as food preferences, is largely a by-product of breeding biology studies and this may sometimes give a distorted picture of such subjects.

To quote two examples, I see the pair of Crowned Eagles living in the forest opposite my house almost daily but although I have an intimate knowledge of their breeding habits and a good idea of their food from what they bring to the nest, I have never seen one kill anything and unless they are soaring in display or near the nest, I can seldom locate them. The Black Sparrowhawk is a very widespread African forest accipiter but, like most accipiters, is secretive and elusive. In sixteen years, I have seen only four or five away from the nest in this forest but for the last four years, a pair has nested not far from my house and, as I write, have two large young. The bird is not shy, just elusive, and will let me approach within 3 metres of her. Yet almost everything I know about this species has been learned here in about 100 metres square.

Egg collectors were the first to study breeding biology. They amassed enormous collections in the late 1800s and early 1900s but unfortunately did little but collect the eggs as trophies. Few made really accurate notes on the construction of nests, clutch sizes, egg weights, and other such details though sometimes these can be extracted from old accounts. In some countries, the study of breeding biology more or less ceased at the end of the egg collecting era at about the outbreak of World War II apart from a few aberrant characters. Thus, in India there are abundant but in some ways inadequate data on nests and eggs but since 1945 there have been very few good details published about the breeding biology of any Indian raptor.

Bird photography at the nest followed and at first, this also was a form of collection or trophy hunting. At least it did not bring the breeding cycle to a summary halt as soon as the eggs were taken, however, and led to fuller studies of the nesting cycle especially with young. Later still, observers have studied the breeding biology alone, with photography as a secondary objective or just as a tool to illustrate points of detail. The trend now is to study breeding biology without photography because the erection of a hide and the disturbance which results from close-up photography may distort strictly natural behaviour.

A few exceptionally tame raptors will tolerate repeated interference but some are very shy and a few, such as the Bateleur, behave in a quite unaccountable manner. Secret watching from a distance with a notebook gives truer information than photography from a hide at close quarters. Small, relatively cheap, time-lapse cameras, operated by radio or automatic controls are now available and can with minimal disturbance be used to obtain a mass of factual data for later analysis which would be beyond the most energetic observer. The incubation temperature of a Golden Eagle, and other details, have been measured by substituting a false egg with a little transmitter in it for a real one recording the results in a van a kilometre away.

Old timers, like myself, who have done our work the other way, sitting cold, cramped, and exhausted for hours, feel that such modern electronic gadgetry (which I should instantly break anyway) can never fully substitute for personal observation and experience however many pretty graphs and masses of significant figures it may produce. But perhaps we are ornithological dinosaurs and I freely admit that such devices may produce results we dinosaurs have failed to get.

Thus, the technique of study of breeding biology has advanced from simple egg collecting to advanced electronics yet it is surprising how much elementary detail remains unknown. For convenience, I assess our knowledge of breeding biology at five levels:

1 unknown (E) – the nest has never been found at all and any reputed eggs are dubiously identified;

2 little known (D) – one or two nests have been found and the eggs at least described;

3 well known (C) – several or many nests have been found, the eggs described, and some other details of breeding behaviour;

4 very well known (B) – at least one nest has been studied through most of the breeding season and the major details of the share of the sexes, food brought, and development of young are recorded;

5 intimately known (A) – one or more pairs have been studied intensively over several years so that besides the other basic details, good accurate data on breeding rates and survival of young are known.

If the knowledge of any species does not fit neatly into any category it can be given an A+ or B– and so on. Though any such assessment is subjective, it helps to clear one's thoughts, and pinpoint research needs.

If this is done, it appears that sixty-two out of 287 species are still virtually unknown and another forty-eight little known; little or nothing is yet known about breeding in two-fifths of the world's birds of prey. I would assess only twenty (less than a tenth) as intimately known and another forty-six as very well known. Less than a quarter of the world's birds of prey are very well known or better in their breeding habits. About 111 fall into the 'well known' category. These vary from species with nests that have only been described three or four times to others in which egg collectors amassed hundreds of clutches but never bothered to learn much more about the birds they robbed.

The unknown species includes one American turkey vulture, *Cathartes melambrotus*, one cuckoo falcon or baza, the Cayenne Kite, *Leptodon*, three eastern forest honey buzzards (one *Pernis* and two *Henicopernis*), one sea eagle from the Solomons, four eastern and two African snake eagles, twenty-one accipiters and near allies, eight out of nine members of the genus *Leucopternis*, two buteonine eagles, two buteos, four booted eagles, three caracaras, five forest falcons, two falconets, and one falcon. Fifty-seven out of sixty-two are forest birds, twenty-two South American, thirty oriental or Australasian, and five African. The eastern and Australasian species, however, include fifteen accipiters with nests that have never been found and without these, the South American forest species would be the least known and would certainly include the most interesting and unusual forms.

In the little-known group of forty-eight species, however, only twenty-four are forest birds, eleven South American, seven eastern, and six African or other. They include the King Vulture, *Sarcorhamphus*, and the Yellow-headed Vulture, the Hook-billed Kite, Pearl Kite, and one snail kite, the two double-toothed kites, *Harpagus*, the large Australian Square-tailed Kite and Black-breasted Buzzard Kite (which may be better known than it seems), a snake eagle, a harrier hawk, two harriers, ten accipiters, one *Leucopternis* and the Fishing Buzzard, *Busarellus*, three buteos and one buteonine eagle, four hawk-eagles of the genus *Spizaetus*, one caracara, the Laughing Falcon, one falconet, and six falcons.

Much more may be known about some of these species than is recorded but I have based this assessment on standard works and recent literature. In some cases, the lack of knowledge is explained easily by the rarity of the bird or the difficulty of observation; for instance, in the forest falcons, *Micrastur*, and some eastern hawk-eagles. In others, such as in those caracaras of the genus *Phalcoboenus* with nests that have never been well described, there should be no difficulty in greatly advancing our knowledge of breeding habits.

In a good many of these species, too, we can forecast what is likely to happen from knowledge of related species. In the genus *Accipiter*, for instance, three are intimately known, another four are very well known, and eleven well known enough to say that their nesting habits do not greatly differ from the best-known species. When someone eventually finds the nest of the New Guinea Black-mantled Accipiter, *A. melanochlamys*, it will probably be a stick nest just like any other accipiter's nest.

Much the same applies to the unknown harriers and buteos, or the little known kites, honey buzzards, or hawk-eagles. For instance, the nest of Cassin's Hawk-eagle recently discovered was predictably in a big forest tree and hard to reach so that the eggs are still unknown and the species rates D. The most interesting possibilities are among the aberrant Falconidae, the tropical caracaras of the genus *Daptrius*, and the forest falcons, *Micrastur*. They are of special interest because only some members of the Falconidae build their own nests. The caracaras might but the forest falcon would more probably nest in a hollow stump like the Laughing Falcon which itself is little known but gives us a clue.

A few enthusiastic, vigorous, determined, and even sometimes unscrupulous individuals have done more to advance knowledge of breeding behaviour than has any co-ordinated research effort. Some egg collectors positively loved the eggs of birds of prey above all others and went to great lengths to get them. Lately, they have unexpectedly proved of some use, because the hard, thick shells of their old egg collections can be compared with the thin, fragile shells of more recent, pesticide-affected raptor eggs, proving beyond reasonable doubt that some pesticides do cause eggshell thinning and threaten the extinction of certain species. This is, however, no cause for egg collectors to boast because that's not what they took the eggs for. Also, recent electrolytic studies of egg whites, useful as a tool in taxonomic research, have depended very largely on illegal or clandestine egg collecting; one American professor is reported to have paid a fine of 3000 dollars under the United States Lacey Act for this reason.

Most of the more detailed work has been done by a small group of amateur enthusiasts and all the outstanding work is still due to amateurs. At any one time, this group has half a dozen front-runners who, in their day, contribute more knowledge than anyone else. Although enthusiastic individuals have taken us a long way, I feel now that there is need for a more solid, co-ordinated research effort, and for channelling available effort and expertise into those areas and among those species where the greatest advances can be most easily made. Dutch workers, under Professor Karel Voous, have recently adopted something of this approach with some outstanding results.

Here I only wish to summarize briefly our knowledge so

that the gaps can be most easily perceived. The species I have placed in various groups are listed in Appendix I. Research so far has shown that there are certain fundamental differences, genetically based, between the behaviour of different suborders and families of raptors. None of the New World vultures (Cathartidae) builds a nest, all lay their eggs on the ground, in caves, on ledges of cliffs, or in hollow trees. Unknown or little-known cathartids will probably behave just like others. Larger condors and the King Vulture lay one egg, the others one or two. The eggs are white, sometimes nicely marked, and when held up to the light, the inside shell is always yellowish. There are probably other fundamental differences of behaviour between cathartids and other raptors but because they are dirty old vultures, some very common, they have attracted little interest. Only one, the almost extinct California Condor even approximates a Grade A.

In the superfamily Accipitroidea, the sole member of the Pandionidae, the Osprey, anatomically very distinct, is not specialized in its breeding habits. It is intimately known; one pair at Loch Garten in Inverness-shire has been more intensively watched than any pair of birds anywhere in the world. These results, however, have never been fully written up. The Osprey builds a large stick nest in trees, on the ground, or on rocky islands, sometimes on artificial structures such as cartwheels, on poles, or on electricity pylons. It lays two to four eggs which are, externally, so richly marked with red-brown that they resemble falcons' eggs more than those of accipitrine kites. Inside, however, the eggshell is greenish so that the Osprey is included in a superfamily with the Accipitridae. The whole course of the nesting cycle is otherwise much like that of a sea eagle, though the Osprey does have a unique distraction display.

All known members of the Accipitridae, including ground-nesting harriers (for details see page 11 ff) build their own nests. These can vary from the very slight, annually rebuilt structures of snake eagles, cuckoo falcons, and some kites to the huge masses of sticks used for many years in succession by the larger sea eagles and booted eagles. Some of the big, colonial griffon vultures breeding on cliffs make very small stick nests but even the Egyptian Vulture and Lammergeier, which breed in caves or on overhung ledges as do some of the New World vultures, make large or at least substantial stick nests. No accipitrid builds no nest at all but a few, especially kites, often build upon the foundation of another bird's nest. The huge nests of sea eagles and large booted eagles are, in effect, a series of nests built one on top of another, accumulating for years in one place.

The clutch in Accipitridae can vary from invariably one in large vultures and snake eagles to as many as ten in ground-nesting harriers; most big species lay one to three eggs and most small species three to five. In all, the eggs are white, greenish, or bluish, often unmarked but sometimes blotched with brown to a varying degree, rarely, as in the Harrier Hawk, almost covered with rich brown markings. In all species the inside shell shows green.

The Secretary Bird (if it is a falconiform at all) resembles Accipitridae in building its own large stick nest and laying pale bluish-green eggs, with a greenish inside shell. Apart from its habits and some anatomical pecularities, it would be described as a big, terrestrial eagle. Many other details of its nesting behaviour resemble those of large Accipitridae.

The Falconidae is the only family in which some members build their own nests and others do not. The aberrant nest-building members are all South American caracaras and it is not clear just where the nest-building habit stops. So far as is known, *Daptrius* caracaras build nests but good descriptions are lacking. Caracaras of the genera *Phalco-boenus, Polyborus,* and *Milvago* certainly do build nests, often in trees, sometimes on rocks, on the ground, or even on buildings but good, detailed observations seem to be lacking. The nests may be very scanty, a few pieces of dung or sticks when on a cliff ledge perhaps foreshadowing the total abandonment of the nest-building characteristic of the remaining Falconidae.

The Laughing Falcon, so far as is known, makes no nest but lays in a tree hollow like a kestrel; forest falcons may well do the same, but none has ever been found. Falconets and pygmy falcons *(Poliohierax* and *Microhierax)* all breed in other birds' nests, *Microhierax* in the dark hole nests of barbets and woodpeckers. In this they foreshadow many true falcons of the genus *Falco,* all of which breed on cliff ledges, on the ground, or in other birds' nests (very often of corvids), and make no nests of their own. Old accounts, including some by quite reputable ornithologists, state that they do sometimes make their own stick nests but the general agreement today is that they do not. The Australian Brown Hawk, *Falco berigora,* is said to do so with more conviction than most others and because one Australian harrier so far departs from normal harrier habits to breed in trees this falcon could bear watching. As a basic rule, however, a falcon found breeding in a stick nest has not built it. Vestiges of nest-building habits may persist in plucking small bits of vegetation or drawing stones towards the breast. These apply to ground- or cliff-dwelling species, however, and more falcons breed in other birds' nests than on the ground.

Most falcons' eggs are especially handsome, roundish, buff or reddish, heavily and sometimes almost completely covered with red-brown or brick-red markings. The inside shell is reddish or pinkish buff in the nest-building caracaras and true falcons alike. The Pygmy Falcon, which breeds in the dark nest chambers of the Sociable Weaver and other weavers and the eastern falconets, *Microhierax*, which breed in barbet holes, lay white or whitish eggs. In their case, we assume that the need for the rather cryptic or concealing red-brown egg colour of those falcons laying on open ledges has been lost or abandoned just as New and Old World vultures have abandoned their sharp talons.

These are basic genetic differences between the suborders and subfamilies and detailed behavioural studies would doubtless reveal many others. They clearly point to the aberrant South American Falconidae as likely to be the most rewarding group to study, however, if new knowledge and not just fun or excitement is wanted. Caracaras may be uninspiring but there is a good chance of picking up more knowledge about the more inspiring Andean Condor and many common but interesting and little-known Accipitridae at the same time.

Looking at the breeding cycle from another angle, the general sequence of events, a remarkable degree of parallelism emerges among the various groups. In all species, the cycle can be conveniently divided into a number of successive phases, in the following order: 1 display; 2 nests and nest-building (if any); 3 eggs and incubation; 4 the fledging period (a) the development of the young and (b) the behaviour of the parents to the young; 5 the post-fledging period, when the young can fly but still depend on the parents for food. To these we should add statistics on the number of young reared per pair per annum and survival rate of the young to sexual maturity to obtain a good idea of the fundamentals of population dynamics. This we can call: 6 breeding success and survival.

We should begin by distinguishing epigamic (nuptial) display from threat display. Although threat display may also be used at the nest, it is developed in more specialized forms away from the nest. Threat display can be used either in defence or in attack and basically it is designed to make a raptor look bigger and more formidable. In typical threat, the feathers of head, neck, and crest (if any) are erected, the wings partly opened, and the bird may sit back on its tail so freeing the dangerous feet and talons to strike forward. This is the attitude of the heraldic eagle. Essentially the same behaviour is enacted defensively by a downy eaglet when menaced by a human or attacked by a shrike and an adult Crowned Eagle aggressively threatening a monkey approaching too close to the nest.

Sometimes (in the European Buzzard, for example) the bird stands bolt upright and opens its wings when confronting another Buzzard. The large Old World vultures have specialized threat displays used in securing a place at a carcass. In these, the head is lowered, any specialized scapular or neck plumes are raised, the tail is sometimes raised *(Aegypius)*, and the bird may advance, in long bounds or walking, stretching out its feet deliberately towards the others. 'Mantling' over prey, with hackles raised and wings tented over the quarry is another specialized version of the threat display. The postures may be accompanied by loud calls, hissing (in vultures), and so on.

The attitudes inform the intruder, 'Come closer if you dare!' or 'Get out!' and are decidedly daunting at close quarters. Moreover, if the intruder does press closer, a violent strike with one or both taloned feet will follow. The bird means business and can do genuine injury, perhaps not important to a sapient human cleverly armed with a sack to entangle the talons, but a different matter for, say, a fox or a mongoose. A German Shepherd Dog, struck in the face by a Crowned Eagle, first with one and then both feet, had to be destroyed.

Nuptial display in such powerful fliers is often aerial, and inspiring and beautiful to watch. Far more energy is often expended by falcons or eagles in nuptial display than in hunting. In New World vultures, aerial displays, if any, do not seem to be spectacular. Some soaring about near breeding colonies is described, with diving and chasing, but this needs more critical study. Most species perform ritualized displays on the ground, near the nest site, however, in which the bright colours of the head and neck, sometimes accentuated, are exposed. The Black Vulture, the only species lacking any bright colours, is the one said to perform aerial displays.

Several New World vultures have bred in captivity and then perform displays presumably similar to those they would perform in the wild. The King Vulture pair run rather rapidly round each other flapping their wings and emitting whistling calls. Wild California Condor males display by facing the mate, opening and dragging the wings, and bowing to hold the bill close to the bright yellowish patch of bare skin on the lower neck. The bird then turns slowly, swaying somewhat, exposing the white underwing coverts of the full adult. This behaviour is clearly like that of the Andean Condor in which the male spreads his wings and raises himself almost vertically, neck curved and slightly swollen, with his bill almost touching his chest. He turns slowly, displaying the white shoulders, and clicks his tongue against his bill to utter a dull 'tok-tok-tok'. The bare, reddish neck skin becomes bright yellow and the whole performance ends with a kind of snorting sigh. These ground displays are unlike anything seen in Old World vultures and suggest that the nuptial displays of cathartids differ fundamentally from those of other raptors.

In the Osprey, the large and numerous family, Accipitridae, and even the Secretary Bird, there are some basic similarities in display patterns common to all with some specialities in individual species or genera. To summarize,

Top left Fig. 31 Threat flight of European Snake Eagle; the bird flies in a somewhat raised posture with slow, deliberate wingbeats.

Top right Fig. 32 Threat display of an injured Crowned Eagle on the ground; wings spread, crest raised, glaring eyes, open beak, feet thrust forward to strike.

Centre right Fig. 33 Solicitation: a nearly fledged Golden Eagle solicits the adult by assuming horizontal posture with wings lowered, often calling loudly; it looks threatening but is begging.

Above Fig. 34 Intruder reaction; the incubating bird lies almost flat, barely visible, and cautiously raises head to ascertain if the human intruder has gone (European Snake Eagle).

Right Fig. 35 Montagu's Harrier in mutual soaring and foot-touching display: (a) male soars level above female; (b) male dives towards female still soaring level; (c) female turns over and presents claws to male; (d) male swings up again, female rolls back and soars level again.

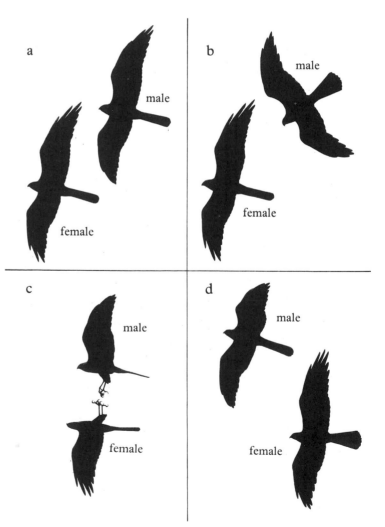

158

Top Fig. 36 Undulating display: shallow undulations. Bird glides then rises with a few wing flaps and repeats (Sparrowhawk).

Above Fig. 37 Typical undulating display. Bird dives with wings partly closed then regains height with vigorous wing flapping. This is often repeated many times (Hen Harrier).

Left : Fig. 38 Undulating display: 'pot hooks'. A series of dives and upward swoops; no wing flapping, the wings remain partly closed (Tawny Eagle).

Below Fig. 39 Undulating display: 'pendulum'. Bird dives, regains height with a few flaps, then swings over and retraces its course repeatedly, sometimes rolling (Verreaux's Eagle).

Fig. 40 'Parachuting'. A male
Augur Buzzard descends
gracefully, like a parachute,
towards the female soaring level
below him; she may then roll
over and present her claws.

these displays occur in approximately the following ascending sequence of intensity:

1 perching and calling – the male (sometimes also the female) perches and calls loudly;

2 soaring and calling – the male, or both sexes, soar over the proposed breeding site, calling at intervals;

3 undulating displays in which one, usually the male, but sometimes also the female dips or dives in flight turning up again sharply and regaining the original height before repeating the performance – this often follows soaring and calling, and may itself be accompanied by loud calling;

4 mutual soaring and undulating displays in which the male often soars above the female and dives towards her, calling or not, and swings up again repeatedly – she may just soar or dive and swoop too;

5 rolling and foot touching in which the male dives towards the female but, instead of swinging up, continues so that she must roll over and present her claws to his;

6 cartwheeling or spinning in which the pair at such a contact lock claws and come whirling down over and over each other, or more rarely, spinning laterally like light, falling leaves.

It seems to me that this sequence of increasing intensity develops logically; for instance, an undulating display cannot be performed until a bird has soared to a height, while cartwheeling or spinning can only be a spectacular development of the mutual rolling and foot touching display.

Besides these, there are certain peculiar displays. The male Osprey soars high above the nest site, flaps his wings vigorously with the body held at a steep upward angle, often dangling the feet and carrying a fish or a bone. It seems as if he can only fly with difficulty and he sometimes travels backwards; he accompanies the performance with a frenzied 'chereeek-chereeeek' call. This display is also performed as a distraction to enemies near the nest and is one of the few raptor displays which have a true distraction function.

In the strongly dimorphic harriers, males are not only smaller, but also very differently coloured, grey where females are brown. They are also often polygamous. At the highest intensity of his nuptial display the male harrier, soaring 30 to 90 metres up, plunges vertically downwards, spinning crazily from side to side sometimes through 360 degrees and uttering loud calls like alarm cries. Spinning and threshing, he seems totally out of control but a metre or so above the ground recovers neatly, mounts again, and may repeat the performance up to a hundred times. In the Hen Harrier, this display coincides with nest site selection. I have once seen something similar done by a chanting goshawk, perhaps allied to harriers, but it is certainly infrequent in these.

Not all raptors perform all these phases of nuptial display

or in this sequence, and in some, voice is more important during or without spectacular flight manoeuvres than in others. In the African Fish Eagle, loud, frequent calling from perches or soaring flight is typical, whether by isolated pairs that never see another or by pairs nesting close together; it corresponds to song. Voice is apparently more important in forest or woodland species than in open-country species. Thus, the Crowned Eagle is extremely vocal while the related Martial Eagle is almost silent. The Laughing Falcon in tropical America is also very noisy. On the other hand, almost all secretive, forest accipiters are very silent and do not perform very spectacular aerial displays above the forest either.

Of the aerial evolutions, many species do no more than soar and call. Accipiters perform shallow, undulating flights while circling. Golden and Verreaux's Eagles perform spectacular 150 metre vertical dives and upward swoops, but usually silently. Crowned Eagles dive and swoop up again accompanied by loud, continuous calling audible far away. Most Old World vultures do nothing more vigorous than a sedate, mutual circling flight, wingtip to wingtip, near the nest site, but the lighter, more agile Egyptian Vulture and Lammergeier apparently perform much more vigorous undulating or diving displays. The ultimate, spectacular whirling or cartwheeling is rare, most often performed by sea eagles and kites.

Fig. 41 'Whirling'. Two African Fish Eagles grapple claws after presenting, and either tumble in a series of cartwheels or spin laterally like falling leaves.

The Secretary Bird performs undulating displays with loud, groaning calls resembling those of other Accipitridae. It also performs peculiar dances and chases on the ground, however, with wings spread, vaguely resembling the dances of some cranes. A mainly terrestrial raptor might be expected to evolve some sort of terrestrial display but these have not been well studied. Again, a comparative behaviour study with cariamas would be helpful.

The nuptial displays of many Falconidae basically resemble the aerial displays of Accipitridae, soaring above the nest site, soaring together, diving at one another, and rolling and foot touching. In the large, swift falcons, the pair indulge in much high-speed, aerial chasing performing evolutions breathtaking to behold. Falcons, however, perform specialized food-presentation ceremonies at or near the nest when bright plumage colours may be important. Hobbies, for instance, sidle along a branch towards each other, the males holding a bird, showing off their reddish belly and thighs, uttering a subdued 'wer-wer-wer'. Male Peregrines feed the females copiously during courtship, bowing up and down and uttering a chittering call. Such food presentation ceremonies do not seem to be typical of equally fierce members of the family Accipitridae and may represent another fundamental difference of behaviour; they need more detailed study.

Display is followed by mating, usually on or near the nest, but sometimes some distance away. In all carefully observed species, mating occurs many times daily, at least for a short period. In the British Osprey, for instance, mating begins the day the female arrives, increases to about seven times daily before egg laying, but ceases just after the eggs are laid. Similar, very frequent mating has been observed in the Goshawk but most species probably mate less often. In the tropics, mating is not confined to the breeding season. African Fish Eagles are likely to copulate almost any day of the year and perhaps frequent amorous behaviour helps to maintain a lifelong pair bond.

In copulation, the male jumps on to the female's back, or he may gracefully alight there from flight. She often solicits him by bending forward until her body is horizontal, partly opening her wings, and calling. She may either raise her tail, or, in long-tailed species such as the Crowned Eagle, twist it sideways to allow union of his cloaca with hers. Mating, apparently successful, can occur in tropical species months, even a year before egg laying but in temperate or cold countries usually happens just before egg laying.

Nothing of substance is known about how caracaras make their nests, and falcons and New World vultures do not make any so that we need consider only the Osprey, Accipitridae, and the Secretary Bird in respect of nest building. Nests of birds of prey are usually built in trees, sometimes on cliff ledges, and rarely (in species that have learned to live with human beings) on buildings or artifical structures such as electricity pylons. When built on cliffs, the nest already has a solid foundation, but when built on trees this has to be made by the birds. Thus, on the whole, tree nests tend to be bigger than those built by the same species on cliffs.

Nests may be just very small, thin platforms of twigs and leaves used for one season only or, at the other extreme. enormous piles of sticks taller than a man and wide enough to go to bed in. In very broad terms, the bigger the raptor, the larger the nest it builds and the longer that nest is occupied. Small kites, honey buzzards, and the smallest accipiters make small, slight nests used for one season or at most two. The biggest nests of the largest sea eagles and of large booted eagles such as the Golden, Martial, or Crowned Eagles are often 2 metres deep and 1·5 metres across and the largest Bald and Golden Eagles' nests recorded are up to 6 metres deep and 2 metres across. The biggest nests tend to be in temperate climates probably because the sticks rot more slowly than in the tropics. The Osprey makes an enormous stick nest, bigger than those of larger eagles. The Secretary Bird builds a huge, flattish, stick nest lined with grass on top of low, thorny trees, much resembling that of the Lappet-faced Vulture.

This basic generalization may be repeated within a large variable genus. Small accipiters, such as the tiny African Little Sparrowhawk, the European Sparrowhawk, and Sharp-shinned Hawk normally build a new nest annually. Goshawks and the large African Black Sparrowhawk may use a nest for several years in succession, eventually building another nearby. New nests are not normally connected with the death of one individual of a pair, or even the female, because when nests are used for many years a succession of different males and females use the same nest.

Many of the larger species have more than one nest and they may alternate between several or use one year after year, sometimes repairing but not using the others. The Golden Eagle may have one to fourteen nests, usually two or three with one preferred. Tropical species usually have fewer nests than related temperate species. Verreaux's Eagle, closely comparable to the Golden Eagle, averages 1·4 nests per pair, while the Golden Eagle in Scotland averages 2·3 to 2·6. The African Fish Eagle averages 1·5 nests per pair while the much larger European Sea Eagle averages 2·6. This may be partly due to quicker rotting and collapse of unused tropical nests. A few large eagles, notably the Crowned Eagle, have only one nest which is used for many years by a succession of different birds.

Exceptions to these general trends occur among colonial Old World griffon vultures breeding on cliffs, which build very small nests for their size but use the same niche of cliff for many successive years. All snake eagles (much larger, for instance than buteos) build very small, slight nests used for one year only, then abandoned. The following year the bird may nest kilometres away or close by and many years later the same suitable tree may be used again. Ground-nesting harriers regularly build new nests each year but usually quite close to the previous year's nest.

Not enough detail is known about the share of the sexes in building but usually both take some part. In fish and sea eagles the male may do more, but in most species the female builds most. She often builds alone but when the male builds, the female is usually present too. The male Goshawk alone is said to build the entire nest in one account but this would differ from any other accipiter known. In polygamous harriers, the female builds most of the nest, the male bringing some material but spending little time building.

Nest building may continue for many months, especially in tropical species where climate is not a limiting factor, or may be compressed into a very short period. In the Honey Buzzard and European Snake Eagle, both completely migrant in their northern range and dependent on specialized food supplies only available in warm weather, no time is wasted in nest building. Broadly, the colder the climate the more compressed the nest-building period for reasons which seem obvious. African Fish Eagles near the equator can visit their nests at almost any season, add the odd stick, or simply stand there doing nothing, but a Bald Eagle in the Aleutian Islands cannot dally in this way. The Scottish Golden Eagle may add some material to the nest in most months, however, though building in earnest may be concentrated in February and March. In such large species, resident in their home ranges year-round, nest building may, like display, help to maintain a lifelong pair bond.

In temperate or cold climates, new nests must be built quickly, but building can occupy many months in tropical species and a nest may not be used at all in the year it is built. When, as is usual in large species, an old nest is repaired rather than a new one built, nest repair may be short and slight or extensive and protracted. Individuals in the same area vary greatly in the amount of material they add, some adding only a few sticks or sprays of greenstuff before they lay, others building 30 centimetres or more of new sticks and lining.

At the end of nest building, fresh green twigs or sprays, sometimes brought from far away but usually collected close to the nest, are added. These continue to be brought almost throughout the nesting cycle and have more than one function, such as part of a courtship ritual. At this stage, however, they mainly function as a relatively soft lining to the nest cup, pressed down by the breast of the female. It does not follow that a nest with green lining will be laid in but eggs are very seldom laid in nests without any green branches. In most raptors, some pairs present in their territories do not lay and we call these non-breeding pairs. The reasons are obscure but such pairs are commoner in the tropics than in temperate climates. In Scotland about one pair of Golden Eagles in five does not breed but about a third of Rhodesian Verreaux's or African Fish Eagles do not lay.

Most birds of prey breed solitarily in their own home range which may be more or less vigorously defended against others of their kind and is at least distinct. Thus, the nests of any species are dispersed more or less evenly over the landscape; this applies not only to species which build nests, such as accipiters, buteos, and eagles, but also to large falcons such as the Peregrine, which make no nests. Several species are, however, colonial breeders, with many pairs nesting close together, and these show no real territorial behaviour.

Truly colonial species, that is, species which seldom nest singly, include the strongly nomadic Letter-winged Kite, the African Swallow-tailed Kite, the Snail Kite, the Plumbeous Kite, and the Mississippi Kite. The Andean Condor and American Black Vulture are partly colonial and all the Old World griffon vultures (seven *Gyps* species) are wholly or mainly colonial breeders, especially the five species which normally breed on rocky cliffs. The tree-nesting Indian and African White-backed Vultures quite often breed singly, but normally breed in groups or colonies. Several harriers are at least semicolonial, pairs breeding closer together than could occur entirely at random; this may be associated with polygamy. Four falcons, the Lesser Kestrel, Red-footed Falcon, Eleonora's, and the Sooty Falcon all breed in colonies. In some areas, notably in the steppes of Asia, the Common Kestrel breeds in colonies and the eastern Red-footed Falcon is perhaps less strongly colonial than its western race.

Some other kites of the genus *Elanus* and the American Swallow-tailed Kite at least breed in rather loose groups that might be called colonies. The huge majority of birds of prey is made up of solitary, not colonial breeders, nesting in a defined home range or territories more or less strongly defended. A few, such as the African Fish Eagle, have been erroneously regarded as at least semicolonial breeders. This species is actually intensely territorial, however, and such cases are just unusually dense populations where pairs can nest very close to one another.

The colonial breeders are mainly insectivorous, carrion eating, or nomadic. Three of the colonial American kites and the African Swallow-tailed Kite are aerial feeders on insects. The Lesser Kestrel and Red-footed Falcon are largely insectivorous and the colonial Snail Kite feeds on snails. The elanine kites, which feed on rats and mice as well as some insects, are all strongly or partly nomadic and the Common Kestrel, where it breeds colonially, is also nomadic. The large colonial vultures are all griffons which feed in numbers together on a carcass and do not defend a territory except for about a metre round the nest site. This probably also applies to the American Black Vulture and Andean Condor where these breed colonially. Harriers are unique in their polygamous, ground-nesting habits but they are also nomadic in their winter quarters. Careful mapping of nesting sites might show that some partly nomadic buteos, such as the Asian Upland Buzzard and Rough-legged Buzzard, also sometimes nested semicolonially where food supplies are temporarily abundant.

Eleonora's and the Sooty Falcon are apparent exceptions; they are specialized feeders in the breeding season on the autumn rush of migrant birds out of Eurasia, and are insectivorous in their winter range. In fact, they share the character of other nomadic, insectivorous, colonial falcons, however, in that they concentrate in the breeding season at a source of abundant food supply. This just happens to be a reliable flow of migrant birds, not an erratic fluctuating population of insects or mice.

Nomadism, carrion eating, and colonial breeding seem to be correlated. Not all carrion eaters are colonial breeders, however, because the American Turkey Vulture, the Egyptian Vulture and Lammergeier, and the powerful Old World vultures of the genus *Aegypius* are all solitary. Even these, however, and such birds as the Black Kite may be less intensely territorial than, say, a Peregrine Falcon or a Common Buzzard.

The eggs of most birds of prey are rather rounded ovals. Falcons' eggs are usually rounder than those of the Osprey and Accipitridae, and the Secretary Bird lays rather long, oval eggs. Though the eggs are sometimes rather large for the size of the bird, they are relatively much smaller, for instance, than those of Kiwis, or even of some ducks. Moreover, because small clutches are the rule rather than the exception, the laying of a clutch of eggs places relatively little strain on females.

In species where the clutches are known, only twenty-four regularly lay one egg, forty-eight lay one to two eggs, sixty-one lay two to three, twenty-five lay three to four, and only twenty-five regularly four or more. Nearly three-quarters of all the known species lay three eggs or fewer, and because more of the unknown species probably lay small rather than large clutches, perhaps four out of five of all raptors lay less than three eggs. Moreover, in most species, the eggs are not laid on consecutive days and in large species such as the bigger eagles, there may be a gap of three or four days between the first and second eggs. In large eagles, individual egg weight varies from about 2·5 per cent to 7 per cent of bodyweight and that of a complete clutch from 2·5 to at most 9 per cent, laid over about a week.

During the whole breeding cycle there are three stages which I call 'points of strain' at which acute shortage of food might cause difficulty. The first of these is at egg laying. It is difficult to believe that in large species, however, shortage of food would prevent a female from laying eggs. In smaller species such as harriers or, for instance, the Rough-legged Buzzard and some falcons, which lay clutches of five or more in some years, egg laying could cause a greater strain on the female. A female Hen Harrier laying a normal clutch of six or seven (not uncommon) might deposit 40 per cent of her own weight or even more. However, neither in the Hen Harrier nor in the European Kestrel in Holland has any connection between abundance of food and clutch size been established, suggesting that food supply at this stage is unimportant. This might not be true of the Rough-legged Buzzard which is perhaps more dependent on fluctuating rodent food supply than any other raptor. In the German Common Buzzard, clutches and broods are both larger in good vole years than in years when food is short.

Again, nomadic, insectivorous, or migratory, but not carrion-eating species tend to lay larger clutches. The elanine kites, nomadic rat and insect eaters, lay four to six and the insectivorous African Swallow-tailed Kite (little known) lays four, whereas the tiny Pearl Kite and cuckoo falcons of about the same size lay two to three. The migratory, aerial, insectivorous Mississippi and Plumbeous Kites, however, also lay one to three so that any link between insectivorous migrants and small clutches is weak; it is stronger in nomadic rat eaters. Harriers, mainly migratory and nomadic in their winter quarters, feeding largely on small mammals and insects, lay abnormally large clutches for their size, regularly four to six, even as many as ten. The unique, tree-breeding Spotted Harrier, *Circus assimilis*, is, however, reported to lay two to three eggs so that perhaps large clutches in this genus are connected only with ground nesting. Falcons as a group tend to lay larger clutches than others of their weight, rarely less than three, often four or five. They also breed on the ground, on cliff ledges, or in other birds' nests; perhaps ground breeding was the ancestral habit and they only took to laying in other birds' nests later. Among buteos, those laying the largest clutches are species likely to be nomadic or migratory that breed in the Arctic or the steppes. Some northern accipiters, such as the Sharp-shin and European Sparrowhawk, which are territorial in breeding quarters and only partly migratory, also lay above average clutches.

Carrion eating, however, is certainly not connected with a large clutch, because all the Old and New World vultures lay one, or at most two eggs. Caracaras and milvagos, as far as known, lay smaller clutches than most true falcons of similar weight. The little information we have about the Laughing Falcon suggests that it lays only one egg. If so, snake eating may be associated with singleton eggs because all snake eagles lay one egg. Large eagles, whether sea and fish eagles, buteonine, or booted eagles, all lay small clutches, usually one to three, averaging less than two, while some, including both very small species such as Ayres' Hawk-eagle, Wahlberg's Eagle, and the huge Martial Eagle, only or regularly lay one.

More accurate comparative data on clutch size are needed, perhaps compiled from egg collectors' registers or the more advanced modern collections of nest record cards. With the exceptions mentioned among nomadic and ground-breeding species, generally, the bigger the raptor the smaller the clutch. Also, tropical species tend to lay smaller clutches than their near relatives in temperate climates. In large species which lay small clutches this is not very apparent; for instance, the average clutch of the tropical Verreaux's Eagle and temperate Golden Eagle is similar, about 1·8 to 1·9. Seven temperate or northern subtropical accipiters average more than four eggs per clutch, while nine tropical or warm subtropical species average less than three. Tropical races of the Common Kestrel, lay three to four eggs, European four to six; 534 British clutches averaged 4·72. This reflects the well-known trend towards smaller clutches in tropical rather than in related temperate birds.

Eggs are usually laid at forty-eight to seventy-two hour intervals, sometimes more, rarely less. Big eagles lay at intervals of three or even four days between the first and second egg; this has a crucial effect on brood survival later. Normally, incubation begins with the first egg but the eggs may not be fully warmed until the full clutch has been laid. Falcons normally begin sitting with the second to last egg. In harriers and accipiters, which lay large clutches but start to sit almost at once, the hatching period is rather shorter than the laying period. A Hen Harrier lays a clutch of four to six over eight to ten days but this may hatch in twenty-four hours to eight days. Accipiters and falcons tend to hatch more nearly simultaneously than do the young of larger species such as buteos or eagles.

In all well-observed species, the female, which sits all night and most of the day incubates most. Only two cases are known (a Merlin and an African Little Sparrowhawk) where the male sat all night, and certainly most records indicate this is abnormal. The male may or may not take a share of incubation by day but is never known to sit for more than half, usually less than a quarter of the daylight hours. Thus, females normally sit for from 75 to 85 per cent

A Male Red-shouldered Hawk incubating in the rain; the female has joined him on the nest, probably anxious because of the rain.

of the total incubation period, often for 100 per cent. Individuals also vary; for instance, the female Crowned Eagle nesting near my house incubated alone but five other female Crowned Eagles all shared incubation with the male. Much observation at several nests is needed to achieve a true idea of the incubation behaviour.

Incubation periods are generally rather long, sometimes among the longest known. The huge condors, and the very large Old World vultures all incubate for more than fifty days while most large eagles sit for forty to forty-five days. Tropical snake eagles in Africa incubate for more than fifty days and the longest recorded period is in the huge Philippine Monkey-eating Eagle, with one record of sixty days. This needs confirmation because in captivity the even larger Harpy Eagle incubated for only fifty-four days.

Although long incubation periods and large size are broadly correlated, the periods are genetically controlled, not by size alone. The African snake eagles and the Bateleur incubate for more than fifty days though they are much smaller than Verreaux's or the Golden Eagle which sit for forty-one to forty-five days. Ayres' Eagles and Wahlberg's Eagle, small species weighing less than 800 to 1000 grams, incubate for forty-five or forty-six days, while buteos of similar weight incubate for only thirty-three to thirty-four days. All falcons, from the tiny, 65 gram Pygmy Falcon to the Gyrfalcon thirty times as heavy, hatch in about twenty-

A dark phase Eleonora's Falcon incubating.

Right Female Hobby incubating
in old Crow's nest; in falcons
females normally incubate and are
fed by males.

Far right Two Swedish Ospreys
on their nest; the male had
brought a part-eaten fish, having
taken a good meal himself first.

Below Males often feed sitting
females; here a male
Sparrowhawk leaves after visiting
his mate.

Below right A unique photograph
of a male Montagu's Harrier at
the nest with the female; males
never normally visit the nests, but
pass food to the female near it.

eight to twenty-nine days. Accipiters, large or small, from the tiny, African Little Sparrowhawk (thirty-one days) to the Goshawk, largest of all, hatch in thirty to thirty-seven days, usually thirty-four or thirty-five. Holstein's statement that the Goshawk incubates for forty-five days is suspect and several recorded incubation periods seem doubtful and need checking.

Where the incubation periods of close relatives are known, the tropical member has a longer incubation period than the temperate or subtropical subspecies. The European Snake Eagle incubates for forty-seven days, the Black-breasted, its tropical race, fifty-three to fifty-five days. Bonelli's Eagle incubates thirty-seven to thirty-nine days, the African Hawk-eagle (slightly smaller and perhaps a good species) forty-one to forty-three days. Such variations are not always clear, however, because Golden, Verreaux's, Imperial, and Steppe/Tawny Eagles all incubate for about forty-one to forty-five days. As a whole, however, tropical species take longer over nest building, lay smaller clutches, and incubate rather longer for their size than do temperate species.

The eggs hatch in the normal way, by the chick cutting its way out with the egg tooth. If the eggs cannot be seen, the imminence of the hatch is signalled by the sitting parent; she or he sits less steadily, sometimes rising to peer expectantly beneath. The hatching chick is audible, squeaking in the egg. Large eagles' eggs have been timed to hatch in twenty-four to forty-eight hours from first chipping to complete emergence, but smaller species hatch more quickly. It is unnecessary and undesirable to disturb birds of prey at the nest at this crucial time simply to obtain such unimportant statistics.

The fledging period can be divided conveniently into three phases of development of the young, which rather closely control the behaviour of the parent as well:

1 the downy phase in which the chick is entirely downy;

2 the feathering phase in which the true feathers grow out through the down;

3 the feathered phase, in which the chick is largely covered with feathers, though traces of down may remain. These phases occupy about one-third, one-quarter to one-fifth, and two-fifths of the whole period, varying from species to species. Young falcons and eagles are not normally called chicks but are given more dignified names, such as eyasses or eaglets. I have never heard anyone speak of a 'vultlet'.

The newly hatched young raptor is helpless and unable to stand, with bleary looking, partly-opened eyes. It can scarcely raise its heavy head on its feeble neck and its body is covered with a thin coating of first down, known as *prepennae* because it grows from the same follicles as the true, pennate feathers. Usually, this first down is thinner and silkier than the second down and is normally one colour, white, grey, or buff, but occasionally patterned. In the Osprey, the brown down is streaked paler so that young lying still in a nest are somewhat concealed; young Martial Eagles are dark and light grey. At this stage, the chick is normally brooded by a parent and the colour probably matters little because conspicuous, pure white chicks left alone in the nest for hours seem to survive. More chicks are conspicuous, white or pale grey than any other colour.

The second down grows after seven to twenty-one days, depending on the size of the young. It is thicker, woollier, presumably warmer, and usually the same colour or a little lighter than the first down. It is known as *preplumulae* because it grows in from the follicles that will later produce the fine, filamentous feathers underlying the adult plumage. Young falcons, at this stage, often look very comical, as if enveloped in voluminous, white fur coats. The young now becomes more active, first shuffling about the nest on its tarsi, but towards the end of the stage learning to stand, back carefully to the edge of the nest to defaecate and, in rare instances, to feed itself on prey lying in the nest. It is still basically helpless, however, and would die if the parents deserted it or were killed.

The true feathers start to come through the down at

Most young eagles at this stage cannot feed themselves, but the young Crowned Eagle, with enormously powerful feet, can tear up prey while still almost fully downy.

various ages from ten days to five or six weeks according to the size of the raptor; in general, the bigger the later. The first to appear are the wing and tail quills, sometimes also the crest, if any. They are followed by contour feathers on the body and usually these sprout all over the body and wings giving the young a spotted appearance, while the head may still be mainly downy. In young snake eagles and the Bateleur, however, the top of the body, head, and wings become completely feathered before the underside. This is apparently a neat adaptation to nesting in a completely open, hot, sunny situation on top of a tree. The Tawny Eagle, which often nests in similar open sites, has also evolved this adaptation, though less fully.

In due course, the young are covered completely with feathers, but the large, specialized flight feathers are not yet fully developed and will continue growing throughout the feathered stage, and even after first flight. During the downy and feathering stages the young have been fed by the parent and cannot normally feed themselves. By the early feathered stage, the bill and taloned feet are almost fully developed, so that when feathered, the young can usually feed themselves on prey brought and are left alone in the nest for long periods, sometimes even days. Most growth now occurs in the big flight quills. Towards the end of the feathered stage, approaching first flight, they climb out on branches, practise wing flapping, and become much more active generally. They may actually lose weight near the end of this stage, which probably helps to make first flight easier. The end of this stage, and of the fledging period, is marked by the young bird's first real sustained flight, perhaps just from one tree to another in a species nesting in forest or woodland, or a bolder flight of several hundred

Three eaglets of the Spanish Imperial Eagle; two are close in age and survived, but the third and smallest disappeared and here has already been rejected though healthy.

Just before leaving the nest young Ferruginous Hawks exercise by bounding up and down.

metres from the nesting cliff in some eagles and falcons. Young Secretary Birds just jump to the ground and may then have difficulty in flying back to the nest.

Adult males are normally smaller (in vultures they are larger) and lighter than adult females, and young in the feathering stage can often be sexed accurately by size and activity. Young males, in my experience, are always more active than young females. Young Sparrowhawks can be sexed by the relative thickness of their shanks, and young Hen Harriers (and probably other harriers too) by their eye colour – eyes of males are greyish, those of females brown. Size and wing measurements are generally the best guide but here we must be careful to note the different ages of the chicks. In harriers and some accipiters, the oldest of a big brood may be completely feathered while the youngest is still downy.

Some young die in the nest, either from starvation or other causes. In some eagles, the main cause is intersibling strife in which the elder eaglet attacks or dominates the younger, preventing it from obtaining food, driving it to the edge of the nest, and sometimes actually pecking at it so that it bleeds and dies of wounds. Conveniently known as the

Female Harrier Hawk at the nest; in front of her the elder chick is killing the younger, but she takes no notice.

Cain and Abel battle, in several species, this battle is invariably fatal to younger eaglets; in others, when two hatch, the second survives in about one case in four. Losses of this sort seem lower in fish and sea eagles than in booted eagles. It is not confined to such large, fierce birds and has been reported in African Harrier Hawks and even Honey Buzzards. Its fatal effects may be speeded by attempts at photography at this stage of the nesting cycle but because it happens anyway in most nests this is probably not very important. The struggle has nothing at all to do with shortage of food because it normally occurs when the nest has plenty of food in it and parents do nothing whatever to interfere. Its evolutionary value is still obscure. Recently, it has been suggested that laying a second egg could be useful as an insurance if the first is lost but analysis of several hundred results from eagles regularly laying one egg or two eggs shows that there is no real difference in the final number of chicks reared per adult pair whether they lay one or two eggs.

In most raptors, including fierce accipiters and large falcons, no such fratricidal strife occurs. In these, broods may be reduced by starvation or shortage of food. In the British Buzzard, it seemed that a shortage of rabbits after myxomatosis not only prevented breeding, but resulted in fewer young reared. This was not confirmed by a detailed analysis of hundreds of British Trust for Ornithology nest record cards. In Germany, Buzzard broods are larger in good rodent years than in bad and in the Arctic, Rough-legged Buzzards certainly have more young in good vole

and lemming years. Most brood losses from starvation occur in the earlier stages when the chicks cannot feed themselves. Then, the elder, if hungry, almost invariably begs from the parent first and receives any available food while the younger may then beg unsuccessfully and will starve if no food remains when the elder is sated. Once big enough to feed themselves, the young usually, but not always, survive to first flight.

Throughout the fledging period, the behaviour of the parents is largely controlled by the stage of development of the young. During the incubation period when the female incubates alone, the male frequently, but not always, feeds her at the nest, and his own spells on the eggs are often connected with feeding her. When the chicks hatch (in all cases where this has been accurately observed, from sparrow-hawks to big eagles) the male at once increases his killing rate, killing twice or three times as much prey as in the incubation period. Presumably, he sees the chicks when he visits the nest with prey but it sometimes appears that he knows anyway without seeing them.

In the early downy and second downy stages the young must be brooded by one or other parent most of the time. Usually the female does most or all of the brooding, but sometimes males take more share of brooding than they did in incubation; for instance, in the Golden Eagle. Males of many species, notably accipiters, take no part in brooding or feeding the young at all, and these would die if the female did not feed them. Males may bring prey but will not feed even if the young beg. In some other species, males can and

Small chicks cannot feed themselves but must be gently induced to reach for food held on the tip of the adult's bill. The gentleness of a Golden Eagle is wonderful to watch.

will feed young if the females are not there or rarely even with females present. One adult male Imperial Eagle took over most of these duties from his mate which though fertile, was in immature plumage and presumably inexperienced. It is evidently desirable that a male should know what to do if the female dies but it is more usual for him to concentrate on hunting and for the female to remain on or near the nest till the young are feathered.

The amount of time actually spent brooding by the female can be reduced almost at once; two or three days old young may be left for hours, especially perhaps, in well-shaded nests in warm countries. The female seldom goes far, however, and will normally return at once if rain falls or the sun becomes very hot; either of these, especially hot sun can kill small chicks quickly – which photographers should remember. Brooding time is reduced from over 90 per cent in the first day or two to nights only at the end of the downy stage. Sometimes, downy chicks are not even brooded at night as in one Golden Eagle watched all night by Mrs Seton Gordon. Few observers continue watching after dark, however.

In the feathering stage, the male still continues to bring food while the female remains perched near but not on the nest. Thus, she is on hand to repel any danger that may threaten the chicks, and often does. The female Crowned Eagle that struck me had a chick in the nest, and all Crowned Eagles with large, downy or feathered chicks are apt to be dangerously aggressive. Even quite small species can be very aggressive. I have been struck by accipiters, falcons, kites, and eagles, and attacked by a good many that did not actually strike. I know that such attacks alarmed my dog, a very large and powerful bull terrier not at all afraid of rhinoceros, much more than they did me. It is probably desirable that the female should stay around and at this stage she brings many green branches to the nest. Green branches are brought all through the nesting cycle. At this late stage, bringing green branches is probably a displacement activity helping the female to stay near the nest where she is needed rather than go off and hunt.

The feathering stage of the young is the second 'point of strain' in the breeding cycle because the male is now feeding himself, his mate, and several young, each of which eats about as much or more than an adult. The number he must feed varies from a minimum of three in large eagles to a possible maximum of fifteen or more in large-brooded, polygamous harriers; it is usually four or five – himself, mate, and two or three young. At this stage, the male's ability to catch prey should be stretched to a maximum, and if food is short he may not be able to catch enough. Thus, if any brood losses are to occur they should occur now but, in fact, they usually occur earlier in the fledging period while the chicks are downy and their appetites smaller. Really good documentary evidence that food shortage causes reduction in broods at this second 'point of strain' is scanty. This in turn suggests that the male can normally kill what is needed almost at will and that food supply may be relatively

Female Peregrine Falcons remain on guard at the nest with growing young, and they and the brood are dependent on what the male kills. The young have voluminous coats of down.

Two adult African Hawk-eagles with two nearly mature young; the male (left) has brought prey, which the two young (right) regard in solicitation posture, while the female has arrived although the young no longer need her help. It is very rare to see both parents at the nest at this stage.

unimportant as a limiting factor in breeding success once the young are large.

Once the young are feathered and can be left in the nest alone, most females can take some part in hunting and usually cease to perch near the nest. In some falcons, however, and accipiters (both types of raptors tending to rear larger broods than, for instance, eagles or buteos) males continue to supply all food right up to the end of the nesting cycle. This is true in the European Sparrowhawk and in the Kestrel which may rear broods of four or five, and eat agile birds and easily caught voles respectively. Usually, however, the pressure on the male to provide enough food is eased by assistance from the female after the young is feathered. She may then bring more prey to the nest than he does and some males almost disappear at this stage. In some accipiters and falcons where the female is very much larger than the male and capable on average of killing bigger prey, the active assistance of the female means that she brings a greater total weight of prey to the nest even if she kills somewhat less than the male.

Most raptors are monogamous and many pair for the life of any individual. Harriers, however, and perhaps some accipiters may be polygamous; a male Hen Harrier may have up to seven females at the peak of his powers and often has three or four. Early studies suggested (as would seem likely) that this was disadvantageous because the females had to leave the nest earlier than normal to hunt for their young. Later results from Orkney showed that, in fact, polygamy results in more young reared per adult member of the population, and that the brood size does not vary much even if the female is often away hunting.

At the very end of the fledging period, the parents tend to bring less food but they do not, as some have suggested, coax the young to fly by tantalizing them with food. The young make their first flights independently, usually when the parents are absent. This is scarcely surprising because they are now left alone nearly all day and night though some female adults continue to roost nearby. If nearly mature young are prematurely frightened from the nest by ringers, for instance, they usually survive. Young raptors should be ringed early in the feathered stage when their legs are fully grown, but they cannot possibly fly.

Fledging periods vary from about twenty-five days in the smallest accipiters and falcons to over 120 days in the largest eagles and vultures. The total time in days is often markedly longer in tropical than in temperate species. Tropical Verreaux's Eagles fledge in ninety to ninety-five days, Golden Eagles in sixty-five to eighty days, averaging about seventy. Other large eagles show similar differences If we consider only the total daylight hours available for hunting, fledging periods in the tropics actually are shorter. A Scottish Golden Eagle could hunt for eighteen out of twenty-four hours in May to July, but a Kenya Verreaux's Eagle for not more than ten, probably only eight out of twenty-four hours. The available hunting hours during the whole fledging period are about 1260 for the Golden Eagle and 720, at most 900, for Verreaux's.

Newly flown young cannot fly strongly because their quills are still growing, 'in the blood' as falconers put it. They remain near the nest for some days or weeks and are fed there by their parents. The early part of the post-fledging period is fairly easy to watch in some species but in wooded or forest areas contact with the young often is soon lost and some leave the neighbourhood of the nest earlier than others. Young Honey Buzzards, for instance, must migrate almost as soon as they can fly. Young Wahlberg's Eagles in Kenya stay only ten to fourteen days near the nest.

Easy, regular contact with both adults and young ends soon after first flight, and most available later information depends on piecing together scattered general observations rather than detailed individual studies. Thus, in the Common Buzzard the young, which leave the nest in July, stay in or near the parental territories until August or September but by October a general exodus occurs and the parents reassert their territorial rights. The dispersing young may set up their own winter territories, not necessarily near where they were reared. Young African Fish Eagles remain near the nest place for up to two-and-a-half months but thereafter move away. Young Buzzards and African Fish Eagles may be attacked by adults but this may be just regular territorial behaviour rather than deliberate expulsion of their own young.

Many people believe that adult raptors drive away their own young to make them independent. In the only full case study of a young Verreaux's Eagle the juvenile was driven away after ninety-eight days. In that time it had moved further away from the nest but had not been known to kill for itself. In Kenya Crowned Eagles, however, the young are still fed by the parents at the nest for nine to eleven-and-a-half months after first flight. The young release the parents from parental duties and are not driven away. During the post-fledging period they certainly learn to kill and a day comes when the parents arrive without food, obtain no response from the young to their calls, and so learn that they have gone. If they rear a young to independence in this way they cannot breed in successive years; similar behaviour is suspected in the Harpy Eagle. The advantage of such a protracted post-fledging period may be to produce a stronger eaglet better able to survive the dangerous period between independence and sexual maturity.

Most statistics of breeding success end with the number of young reared to first flight per adult pair. This is as far as we can go in species where contact with the young is soon lost. Various methods of marking young, by cutting windows in wings and tail, dyeing them, or ringing them have been tried, but with little long-term success. Probably not much more will be learned until radio tagging and tracking methods have been further developed.

For accurate statistics of survival up to the stage of first flight three basic figures are needed:

1 the total number of pairs in the area – far too often, only breeding pairs are counted;

2 the number of pairs that breed – in large species and in tropical species, a higher proportion does not breed than in small species or those of cold climates, though as mentioned, fluctuating food supplies can also affect breeding in Arctic Rough-legged Buzzards and Gyrfalcons.

3 The number of young leaving the nests. From 3 divided by 1 we can calculate the number of young reared per pair or per adult overall which is the fundamentally important figure. If only the pairs which breed, or those nests which succeed in rearing young are counted, the result is not only inaccurate but highly misleading.

The available good figures of this sort show that tropical species have lower overall breeding rates than do temperate species. Thus, the British Golden Eagle produces about

0·83 young per pair per annum overall if human interference is discounted; human interference reduces this figure to 0·56 per pair. Unmolested Rhodesian Verreaux's Eagles, however, produce only 0·51 young per pair per annum and African Fish Eagles at Lake Naivasha produce only 0·47 young per pair per annum compared to about 1·02 young per pair in Alaskan Bald Eagles – more than twice as productive. Small species such as accipiters and kestrels have not been studied in the tropics as they have in temperate climates so that comparison is impossible, but because they generally lay smaller clutches, breeding success is likely to be lower too.

In the tropics, in general, fewer pairs breed, those that breed lay smaller clutches of eggs, and fewer young per pair overall are reared. If survival to sexual maturity post-fledging is similar, it follows that tropical species must be longer lived than their temperate counterparts. All large-scale ringing has so far been done in temperate countries, however, so that even the data on survival to maturity derived from ringing, suspect though they may be, are not available for tropical species.

Ringing records for temperate species all show that high mortality of recently independent young in the first winter makes this a crucial period. It is, in fact, the third 'point of strain' of the breeding cycle, which must, in the end, have evolved to produce enough adult mate replacements, one to several years later, to maintain a stable adult population. In any stable population, breeding birds are all full adults and if established pairs include many subadults or immatures, the species as a whole is likely to be in trouble.

The effect of losses in different stages of the breeding cycle on species survival cannot be even approximately calculated for more than a few temperate and tropical species because the data are just not available. The effect of different points of strain must vary from species to species, however, and is not uniform. It may be that egg losses, the Cain and Abel battle, brood reduction from starvation, or, as shown recently with large African vultures, the stage just after first flight when the immatures must compete with the adults for food, is most important. But the breeding cycle must often begin some months before the crucial stage for survival of the population as a whole and how do the adults know when to start?

After this very sketchy and generalized survey, those specially interested in the breeding cycle must be referred to the voluminous literature on many species. We may remember, however, that little or nothing is still known about two-fifths of the world's birds of prey, so that there is plenty of research still needed at an elementary level. Even in the twenty-odd species which are the best known, the last word on their breeding behaviour has certainly not been said.

The ecology of predation

The complex interactions between predators and their prey is a subject which has fascinated and largely defeated a number of very able scientists. Books have been written about it and no subject is likely to arouse hotter controversy. Perhaps this is partly because man is, or was, a predator himself, or at least because predators do at times affect his interests. No full understanding of the subject is at present possible in diurnal birds of prey and here I can only briefly summarize the basic methods by which we can arrive at a partial understanding of the difficult problems involved.

This book is about diurnal predators and strictly nocturnal predators such as owls need not concern us. In fact, many nocturnal animals are stirring by dusk and some diurnal predators do feed on nocturnal creatures. For instance, the Bat Hawk feeds almost exclusively on bats. Studies done on nocturnal predators, however, indicate that many of the same principles apply as by day.

Almost all diurnal raptors feed on relatively small animals. Those that do not are carrion-eating vultures and are a special case to be discussed separately. Again, we can eliminate a large number of animals as potential food for

Locusts formerly provided abundant food in huge swarms, but are nowadays largely controlled; grasshoppers are still plentiful and many raptors eat them.

diurnal birds of prey. Even within the edible size range, however, we have raptors, large and small, that eat everything alive from termites and grasshoppers to crabs, snails, frogs, fish, snakes, lizards, birds, and mammals weighing up to 5 kilograms, occasionally larger. In a great many of these prey species such as snakes or snails, no attempt has been made, so far as I know, to correlate the relationship between predator and prey except in a very general way. More detailed studies have been done on predators that eat mammals and birds than on any others and I shall concentrate on these studies.

The Peregrine Falcon is known to kill more than 120 species of birds, and clearly here the problem of correlating predation and prey is very complex. With mammals, however, the relationship sometimes appears relatively simple; for instance, a buteo or a kestrel feeding almost entirely on small rodents, or a large eagle preferring one type of medium-sized mammal. The simplest predator–prey relationship I know is that between Verreaux's Eagle and hyrax; 99·9 per cent of this eagle's prey is hyrax. The Rough-legged Buzzard, both in winter and in summer breeding quarters, feeds mainly (90 to 95 per cent) on small rodents of two or three species so that these cases appear simple.

While it is easy to count Verreaux's Eagles, however, it is extremely difficult to count the hyrax living among piles of huge boulders overgrown with vegetation, partly nocturnal, and highly mobile. The populations of rodents eaten by the Rough-legged Buzzard themselves fluctuate cyclically over periods of years and the Buzzard breeds

Rock Hyrax; rabbit-sized mammals related to elephants, they are the staple prey of Verreaux's Eagle, whose range is coincident with theirs.

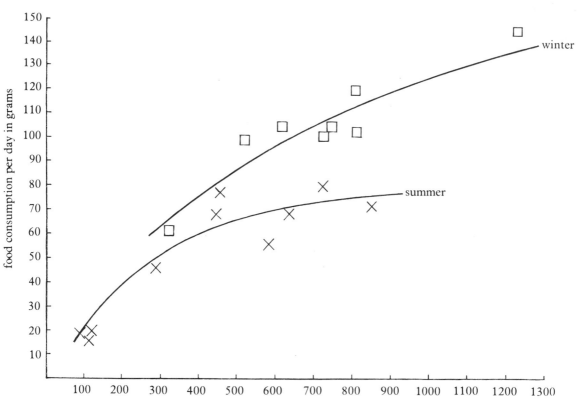

Fig. 42 Food needs of raptors, winter and summer. The amount of food eaten does not increase in direct proportion to bodyweight, larger species requiring relatively less, irrespective of season.

Fig. 43 Diagram of the food needs of birds of prey; the basic need is increased by appetite or cold weather to some extent, and a proportion of each kill is wasted, so that the amount killed is considerably more than the basic need.

actual amount killed to obtain basic needs
wasted portions of kills

increment due to exercise
increment in cold weather

basic minimum requirement

more often, lays bigger clutches, and rears more young in good rodent years than in bad. Thus, even in the simplest cases, the relationship is more complicated than it seems.

A further difficulty is the number of other predators eating the same thing. Hyrax, for instance, are preyed upon by mammalian carnivores from leopards to mongooses, at least five other large diurnal raptors, and large owls, and, in the crevices between the boulders, by puff adders and cobras. Tundra rodents are also eaten by the Snowy Owl, skuas, Ravens, and Arctic Foxes, to name a few. Any complete study of the ecology of predation should, therefore, take all these predators and possible prey into account, and this is why it has defeated most who have attempted it. Though partial solutions are certainly possible, I doubt if there is even one study of predators and their prey entirely satisfactory.

We must start by concentrating on one subject and relating the effect of one possible predator on its prey by the following procedure which, given certain basic data, is a matter of simple arithmetic.

1 Appetite – the amount eaten by a bird of prey is a function of its bodyweight varying from 20 to 25 per cent of bodyweight in small, active species to about 5 per cent in the largest eagles. A raptor cannot eat more than a certain amount any more than you or I can. Like most purely

carnivorous creatures, diurnal birds of prey can, given the opportunity, gorge several times their average daily need. A Golden Eagle can consume 1500 grams at a sitting – about six times its average daily need. It will not then be hungry or inclined to attack anything else for some time, however, though it may take another snack off a part-eaten kill. A large raptor can then go for days, even weeks, without feeding. The average daily appetite can be quite precisely estimated by keeping captive birds of prey for several months, exercising them by falconer's techniques, and weighing them before and after feeding. Appetite is increased 10 to 40 per cent by cold weather and about 10 per cent by exercise but the basic appetite of a resting bird of prey is not doubled or multiplied several times over by such variations. A tropical species should eat proportionately less than an Arctic species, a fast-flapping flier more than a glider, old birds less than young, and so on.

2 The waste factor. To obtain its basic daily needs, a raptor must kill a greater weight of prey than it will eat. The stomachs of large mammals, the fur and some larger bones, and the feathers of birds are rejected or cannot be digested. The waste factor is normally 10 to 30 per cent but can vary from practically nil to over 50 per cent according to type and size of prey. Fish and snake eaters digest everything, bones and all, except the scales but much of a bird is feathers and is wasted, while more than half of a large mammal that must be dismembered may be lost to other predators or abandoned. By adding the amount wasted to the amount eaten (the appetite) we arrive at the total amount killed or used by a raptor daily, or for any other period. In the Scottish Golden Eagle, for instance, the amount killed or needed is estimated at 87 kilograms per adult per year of which 84 kilograms are actually eaten.

3 The numbers of raptors. Most, even migrant species, live in pairs dispersed over their breeding areas and many are permanently resident in a home range year-round. I dislike the word 'territory' in connection with raptors be-because it usually means a defended area and many raptors do not obviously defend the whole or even a large part of the area they occupy. Breeding pairs are normally dispersed rather evenly over the available terrain so that the population of any area can be quite accurately esti-mated. Colonial and nomadic species are obviously special cases that do not fit this general pattern. Usually, we can estimate the total area occupied by a pair, and from stages 1 and 2 the amount they will eat and kill per year within this area, allowing for the difference in size and weight between the smaller males and larger females.

4 The breeding success or replacement rate. The raptors

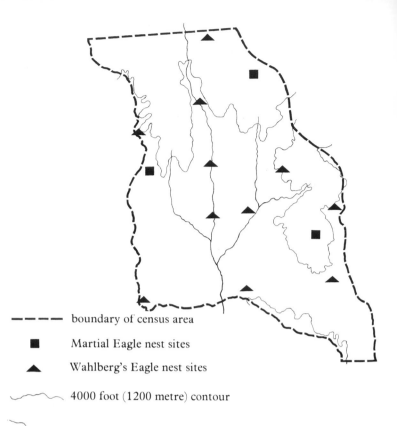

Fig. 44 Map of a census area of 146 square miles (378 square kilometres) in Embu District showing breeding sites of Martial and Wahlberg's Eagle.
Both hunt over the whole savanna area, but each is evenly dispersed, and the territories of three pairs of Martial Eagles weighing up to 6000 grams are much larger than those of Wahlberg's weighing 1000 grams.

- - - - boundary of census area

■   Martial Eagle nest sites

▲   Wahlberg's Eagle nest sites

〰️   4000 foot (1200 metre) contour

〰️   main river in area

breed, more or less annually, and over a period of years produce a certain number of young. These require food in the nest and for at least the period thereafter until they are independent which might vary from a few weeks in the case of large, carrion-eating vultures to eleven-and-a-half months in the Crowned Eagle. Unfortunately, there is no short cut to good statistics of this sort. Either a group of raptors must be studied for many years in succession (the longer the better), or several groups must be studied for several years (which probably gives a more reliable answer), or a very large number of individual pairs, not forgetting non-breeding pairs must be recorded for one or two years. Fifty to 100 pair years, more if possible, is what we should aim at. In a species such as the Rough-legged Buzzard, of which prey fluctuates cyclically, it would be necessary to follow the same group or groups of birds through at least one, preferably two successive cycles of abundance and shortage of the prey to arrive at a sound figure. So far as I know, this has never been done.

5 The total demand. Knowing the number of adults and the number of young they produce annually, we can calculate the total weight of food needed by the raptors and their progeny from the home range. Immatures and sub-adults or unattached adults must also live somewhere, and often must live in the ranges of established adult pairs.

Sometimes the number of such immatures can be estimated if they are easily recognizable; for instance, in Bateleurs about 30 per cent of the total population are immatures. In the Golden Eagle 20 per cent immatures and surplus adults is probably about right. The total demand on the home range is the weight of food taken by the adult pair, their dependent progeny, and the sub-adult followers necessary to provide new mates to maintain a stable population when adults naturally die.

6 Food preferences. We then must ascertain what these raptors eat and what these animals weigh. Prey preferences can be estimated from prey brought to the nest (generally regarded as the most reliable method though it can be misleading), from kills observed or located (usually difficult), and from castings or pellets found under roosts or near nests. This last method has been extensively used but is rather unreliable because the bones of different animals are not all digested at the same rate. Bones of frogs, reptiles, and fish, even sometimes of mammals, may be completely digested so that what appears in castings may give a biased result. Bias is almost unavoidable in any method, however, and we must do what we can, allowing for it. Knowing the preferred prey and their weight (not always easy in fish or mammals which vary in size according to age), it is then possible to calculate how many of each type of animal must be killed to provide the basic demand of the raptors. For instance, in the Scottish Golden Eagle, the total basic demand of 271 kilograms is provided by 70 hares or 110 rabbits, about 150 large birds (mainly grouse and Ptarmigan), and 50 kilograms of carrion.

7 Availability of prey. We must estimate how many of the prey live in the home range. This is where most studies break down. Very few indeed have accurately correlated the raptors' demands with the numbers of available prey because one man can hardly watch and study the raptors and at the same time study and count their prey. In the simplest known situation, no good method of estimating the numbers of hyrax in a Verreaux's Eagle's home range has yet been devised. Species eating a great variety of prey, or with different diets at different seasons, are evidently much more difficult. For instance, the American Red-shouldered Hawk feeds largely on small rodents in winter but in summer eats many snakes and frogs, and their numbers are extremely hard to estimate. Some situations should be much easier; for instance, in the western United States, the Golden Eagle eats 98 per cent rabbits and jackrabbits (hares), and their numbers in a home range should not be so hard to estimate; again, this has not so far been recorded, to my knowledge.

8 Vulnerability of prey. Prey may be abundant but if the

raptor cannot see it, it cannot catch it, or could only catch incautious individuals which expose themselves unnecessarily. Any fishes that stay a metre below the surface cannot be taken by fish eagles, but are vulnerable to cormorants. In dense Mediterranean *maquis* rabbits could abound, but Bonelli's and Golden Eagles could not catch them though mammals such as ferrets and foxes could.

The vulnerability of the prey can be assessed and given a rating, A, B, C and so on. It may vary according to the habits of the prey, their numbers, or from day to day, even from hour to hour. When a grass fire sweeps across African plains, it first drives out otherwise well-hidden grasshoppers to be eaten by kites and Grasshopper Buzzards. After it has passed, rodents formerly completely screened by long grass, are exposed to attack. A drumming, male Ruffed Grouse, or a crowing cock Pheasant may be far more vulnerable to a raptor than the well-camouflaged, secretive female. A Red Grouse is vulnerable to a Golden Eagle on the ground, and in flight to a Peregrine Falcon living in the same area. A hyrax is only vulnerable to a Verreaux's Eagle when sunning itself on a rock or feeding in the topmost branches of a tree. Flocks of common birds, such as sparrows, are more conspicuous and vulnerable than are skulking, uncommon species of the undergrowth, and so on. An animal can become vulnerable through an unusual situation; a normally well-hidden snake or a rat crossing a tarmac road immediately becomes vulnerable to an Augur Buzzard sitting on a telegraph post – which is why buteos do sit on roadside posts.

Thus, it is plain that to assess all the factors involved in even the simplest situation is no easy matter. The difficulties multiply with the number and variety of prey and predators involved. In the poorest of all habitats, the tundra, the food of the four regularly breeding, diurnal raptors is quite well known, and a concentrated study covering one or two population cycles of the prey animals would produce quite a good estimate of the relation between predator and prey. In tropical savanna or forest, however, with over 100 species in the habitat as a whole, and twenty or twenty-five in any one locality, including some with nests that are still unknown, a similar estimate is totally impossible on our present knowledge.

In my opinion, no one person can even attempt such work in complex situations with a real hope of success. To study predation effectively in, for instance, the 110 square kilometres of the Nairobi National Park would ideally require a plant ecologist, a mammalogist, a herpetologist to study snakes and frogs, and two ornithologists, one to study the raptors themselves and one to study the other birds. All would need to be working simultaneously for, probably, at least three years because some of the raptors such as the Secretary Bird and Black-shouldered Kite are partly nomadic.

The evident difficulties have not deterred several scientists from attempting at least a partial solution, and studies within the last forty years or so have amassed a great many useful facts. To date, no full studies have been attempted in tropical areas but several good studies of individual species, groups, or populations have been done in temperate countries. Although these can give only a partial answer to the complex problem, they do provide useful guidelines. It is a pity that no adequate attempt has been made to study the problem in tropical countries because, in most temperate countries, the environment is no longer completely natural, or the raptors are persecuted, or both.

From results to date, opinion tends to polarize along two lines. The first, or orthodox view, is that raptors must work hard for their food, and their population dynamics and relations with the prey are, therefore, controlled mainly by the abundance of the prey. The other extreme view is that because raptors apparently catch their prey with ease, total food supply alone may not be the most important factor. Social behaviour might be more important in population dynamics and consequently in the effect of the raptor on its prey. I cannot believe, for instance, that the Fish Eagle, which spends only a little more than one hour in a hundred actually hunting, can be controlled by the difficulty of catching food. On the other hand, a male Buzzard on Dartmoor which had to hunt for eleven hours to catch enough voles to feed its brood, would clearly have run into trouble if voles had been less numerous.

Scientists working in cold or temperate climates tend to hold the orthodox view but those working in the tropics may be more inclined to admit that the other view is tenable. The late Professor A J Marshall believed that it was absurd to suppose that one single factor could control everything, and to those who hold rigidly to orthodox dogma, I would repeat Oliver Cromwell's celebrated injunction to his contending advisers, 'I beseech ye, in the bowels of Christ, think it possible ye may be mistaken'.

The best and fullest study done is that of John and Frank Craighead in Michigan and Wyoming. They studied both diurnal and nocturnal raptors and their favoured prey, winter and summer, in 1942–43 and 1947–48; and published the results, very fully supported by tables and basic data, in their book, *Hawks, Owls and Wildlife*, in 1956. This study is by far the fullest yet available in print, and the basic methods they used and developed must be followed by everyone who attempts anything similar, with any possible improvements. This book is absolutely indispensable to anyone who aspires to understand the relation between raptors and their prey.

Most other studies concern a single species, such as L Tinbergen's monumental study of the Sparrowhawk in Holland, or a group of related species such as two or three accipiters, or, more recently W J A Schipper's detailed analysis of the prey and behaviour of the three European harriers. The interrelations of different African vultures have been studied by H Kruuk, and here I can only touch upon the results of some of these studies and refer the interested reader to them for further details. It is also useful to learn about the effects of tigers or wild dogs as a comparison.

The Craigheads studied three buteos, Cooper's Hawk, the Hen Harrier (Marsh Hawk), and American Kestrel (Sparrowhawk) in winter and in summer, and several owls. They also estimated the populations of the main species of prey, small rodents, squirrels, rabbits, game-birds, and other birds, and less accurately frogs, insects, and snakes. They determined prey preferences at different seasons by pellet analysis (a method liable to lead to bias, as they admit) and by direct observations at the nest or location of kills. They could assess quite well the effect of the predators on their prey in their particular conditions. All three buteos, the Hen Harrier, and the American Kestrel, and all the owls studied, fed in winter mainly on the small rodents so that the relationship was relatively simple. A collective population of raptors fed mainly on one type of prey, small mammals.

They concluded, basically, that the number of raptors present in winter depended primarily on the abundance of prey, that is, availability, and secondly on its vulnerability which was increased by numbers leading to greater conspicuousness, and the behaviour of the rodents themselves. Large rodent populations ate down their own vegetative cover, exposing themselves to attack, while a small population of rodents unable to consume their long grass habitat could not become vulnerable in this way. A catastrophe, in this case a flood followed by a frost, filling formerly protective burrows with ice, forced the rodents to the snow-covered surface, to become vulnerable.

In these circumstances, the Craigheads concluded that the raptors did limit the numbers of their prey, reducing high autumn populations to low spring numbers and concentrating to hunt in those areas where the rodents were most numerous and most vulnerable. Apart from the bird-eating Cooper's Hawk, however, all their species were primarily rodent eaters, their diet being 87·2 per cent to 98·9 per cent rodents. The winter population of raptors was also partly migratory and locally nomadic; there was almost four times the number of hawks (213) in autumn and winter combined of 1942–43 as in 1947–48 (56). Thus, the conditions seemed right for a concentration of raptors feeding on a particular food supply and the birds behaved much as we would expect locally nomadic species to behave. The situation was not comparable to one quite commonly seen in the Kenya Rift Valley, for instance, where a stable population of resident raptors is totally unable to control a plague of grass rats running about on the bare ground. Sated raptors then sit around ignoring easily caught prey, and some other factor, probably disease, eventually cuts the rodent population down.

In spring and summer, more birds, reptiles, frogs, and crustaceans were taken by some of the raptors so that the proportion of small rodents taken fell to about 35 per cent. Moreover, the raptor population now became stable, old established nesting territories or ranges being re-occupied. Adding spring and summer together, the total number of raptors was almost the same in both years of study, 241 compared to 238. The pressure exerted upon the prey was no longer wholly or mainly dependent on the density and vulnerability of the prey but more on the dispersed nesting pattern of the raptors themselves. During spring, these continued to depress the small populations of prey left at the end of winter but, to some extent, could turn to other prey not available in winter. Although the maximum demand of the cumulative breeding population (of two buteos, one accipiter, the Kestrel, and several owls) was greatest when they themselves had broods in the nest, they could not prevent the prey increasing, they only reduced its rate of increase.

This summer situation was more like the type of stable resident raptor population normally found in tropical areas when the raptors often appear unable to control or even markedly affect the numbers of prey animals present in its home range and the raptor populations cannot, therefore, be limited by the available prey. Comparable quantitative studies for any tropical area are lacking among birds of prey, however. In the Craigheads' study area, the essence of the matter apparently was that during autumn and winter, a resident population of raptors, augmented by partly nomadic migrants, depressed the high late summer populations, especially of rodents, to a low level reducing its potential for rapid breeding in the spring. Further predation in spring still further reduced the remaining populations but once the raptors had dispersed in nesting territories, the prey had a respite and was able to multiply again. Supposing that the raptors had not been there, the high autumn populations of rodents would have eaten all available food and then starved. There might then have been fewer breeding rodents in spring than there

Right In tropical countries downy eaglets are easily killed by hot sun so that they are often shaded by a parent; a female Verreaux's Eagle here shades her chick.

Below Both parents are not often seen on the nest together; here a male Kestrel brings food to female and brood.

Above As young Common Buzzards grow and develop feathers, they are left alone in the nest by both parents; the female now shares in hunting. Broods of three are seldom reared unless food is abundant.

Left A young buzzard just out of the nest; it can fly but cannot kill its own prey and is dependent on its parents for a further three weeks or more.

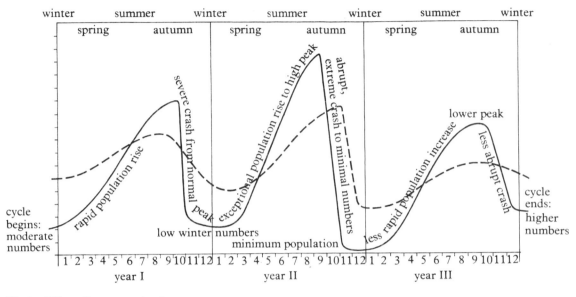

normal eruptive cycles, varying annually, with steep summer increase followed by
——————— abrupt crash to low numbers in winter: crash is worst following exceptional numbers
in year II and population cannot recover so fast in year III
population curve smoothed by raptors: in year II predation cannot prevent a moderate
– – – – crash, but it will be to a lesser extent than if no predators were present

Fig. 45 Effect of raptors on rodent
cycles (diagrammatic) to show
how more extreme eruptive
cycles are partially controlled by
predation.

were with buteos, kestrels, and owls all eating them. The
effect of the raptors was to smooth out the curves of the
eruptive cycles of population the rodents would otherwise
have completed (Fig. 45).

Tinbergen's study of the Sparrowhawk, entirely feeding upon birds, and mainly on small birds weighing less
than 30 grams, also showed that vulnerability to attack was
increased by abundance or conspicuousness. House
Sparrows, very numerous and inclined to gather in conspicuous flocks in the open, were most often taken (hence,
if you were in any doubt, Sparrowhawk), and predation by
the Sparrowhawk accounted for about half the total mortality of Sparrows. Tree Pipits, however, which were not
very numerous but had conspicuous habits of perching
and singing, were taken more often than expected in proportion to their numbers, and were also vulnerable to the
Hobby when in display flight. Swallows, numerous and
conspicuous, flew too fast in the open for the Sparrowhawk to catch them often, and skulking, inconspicuous,
uncommon species were taken less often in proportion to
their total numbers. Thus, in considering the effect of one
possible predator on a large variety of prey, the selective
nature of predation was apparent. The Sparrowhawk may
control Sparrows to some extent, but not Swallows, which

must be controlled by something else, possibly not a predator at all.

Several American studies of grouse and Pheasants have shown that the displaying males are likely to be more vulnerable to, for instance, the Goshawk and Red-tailed Hawk. You have only to look at a brilliant cock Pheasant in the open to see that he could be exceedingly vulnerable. On the other hand, few territory-owning, adult, male Red Grouse are taken by predators which mainly take surplus, unmated or young birds.

We can understand that a cock Pheasant might be naturally expendable once he has mated his hens but a male Tree Pipit, that shares in feeding the young, should not render himself vulnerable by conspicuous display. There must, in fact, be some extraordinary advantage to be gained from display in which males apparently render themselves more vulnerable to predation than they need. In some African plains' mammals, more males are taken than would be expected from their numbers in the total population. Apparently, the need to establish and retain a territory makes them reluctant to leave it, so that they are eaten for their tenacity. Yet such tenacity must benefit the population as a whole or natural selection would eliminate it.

Sheer numbers of the prey sometimes render them almost immune to any real predation effect. The Sudan Black-faced Dioch, *Quelea quelea*, for instance, is not controlled by predatory birds though a large number, from small accipiters to large falcons, eat them, and at breeding colonies of millions of nests, Tawny Eagles can be seen sitting around gorged and unable to eat more of the available superabundance. Egyptian Vultures in pelican and flamingo colonies eat many eggs and break far more than they actually eat but have no important limiting effect on the total numbers of young reared. On the other hand, Andean Condors are said to alight in colonies of Peruvian seabirds and cause mass desertion, doing far more damage than they would by simply scavenging a few dead chicks. I can readily believe this happens, because I have repeatedly seen a few Marabou Storks cause mass desertion of colonies of thousands of pairs of Greater Flamingos. At one extreme, a species can be so numerous that it outstrips all possible predation and at another a few predators can do so much damage through sheer fright that their real predatory effect is negligible.

Where a large variety of predators occurs in the same area, they generally avoid competition for the same sort of prey. They may be graded by size from very large to very small; for instance, among ten Swedish raptors the size range is almost the same as among thirty-nine in Surinam. Thus, they are potentially able to kill very

Right White-backed Vultures, a
single Rüppell's Griffon, and a
Marabou Stork on an elephant
carcass; carrion birds eat the
largest of all mammals when dead.

Below Griffon Vultures may
threaten, but cannot repel even
pi-dogs at a carcass.

Below Adult Sykes Monkey killed by a Crowned Eagle; an hour later the Eagle has eaten well, and dissected and removed one hind leg. Such large prey cannot be carried entire, and may be partly wasted.

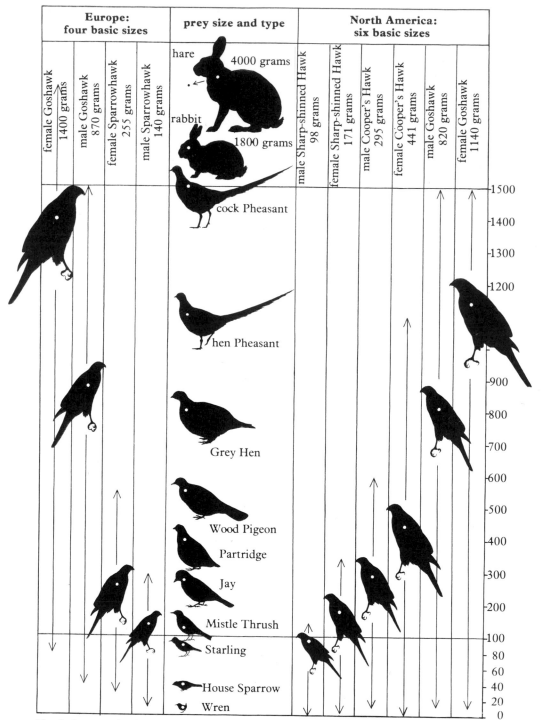

Europe: four basic sizes				prey size and type	North America: six basic sizes					
female Goshawk 1400 grams	male Goshawk 870 grams	female Sparrowhawk 255 grams	male Sparrowhawk 140 grams		male Sharp-shinned Hawk 98 grams	female Sharp-shinned Hawk 171 grams	male Cooper's Hawk 295 grams	female Cooper's Hawk 441 grams	male Goshawk 820 grams	female Goshawk 1140 grams

hare 4000 grams

rabbit 1800 grams

cock Pheasant

hen Pheasant

Grey Hen

Wood Pigeon

Partridge

Jay

Mistle Thrush

Starling

House Sparrow

Wren

weights of accipiters and prey in grams (*see note*)

1500
1400
1300
1200
900
800
700
600
500
400
300
200
100
80
60
40
20
0

Vertical arrows indicate prey overlap. Average weights are indicated by white dots in centres of silhouettes. Silhouettes are approximately to scale.

Fig. 46 Ecological separation among European and North American accipiters.

different-sized prey and some may also have specialized diets, such as snakes, crabs, or snails among some Surinam raptors (less so among the Swedish ones); the variety can be greater for this reason too.

Similar-sized raptors eating the same prey can avoid competition by different habits. Thus, of the two snail kites feeding on snails in Surinam, one is broad winged and short tailed adapted to hunting from perches, the other rather long winged and long tailed catching the same sort of snails over open marshes in flight. On the Andaman Islands, two small snake eagles avoid direct competition for available snakes by inhabiting respectively coastal mangrove forests and interior rainforests.

Where similar-sized species occur, feeding on the same type of prey, they are normally geographically or ecologically separated. Thus, five similar-sized buteos occur in Africa. The Common Buzzard occurs only on the offshore Atlantic islands. Its ecological niche in north-west Africa is filled by the Long-legged Buzzard. In tropical west African savannas the only buteo is the Red-tailed, replaced in more elevated east African savannas by the Augur Buzzard. In South Africa only the Jackal Buzzard occurs over most of the country. In South Africa, and in east Africa and Ethiopia, the mountain forests are inhabited by the Mountain Buzzard, which avoids competing with the Jackal or Augur Buzzards because it is ecologically separated. On Mount Kenya, the Augur Buzzard occurs on the plains below and the heathlands above the forests, the Mountain Buzzard between. This rather neat arrangement apparently breaks down, however, when in winter the flood of migrant Steppe Buzzards pours southwards through Africa to South Africa where they ought to compete for available rats with both Jackal and Mountain Buzzards. They may occupy rather different habitats, however, or the rats may be so abundant that it does not matter. If not, then it is not food supply that controls the number of buzzards.

Possible competition between closely related species has been more thoroughly studied in the genus *Accipiter* than any other. The neatest comparison is between the three American species, the Goshawk, Cooper's Hawk, and the Sharp-shinned Hawk. In all accipiters, males are much smaller than females, so that, in effect, three species produce six different sizes of predators, killing six different sizes of prey. Male Sharp-shins, averaging 98.8 grams themselves, killed prey ranging from 3.4 to 135 grams, averaging 17.7 grams. Females, themselves weighing 171 grams, took prey of 3.4 to 343 grams, averaging 28.4 grams. Male and female Cooper's Hawks, averaging 295 and 441 grams killed prey averaging 37.6 and 50.7 grams. Goshawks averaging 808 and 1137 grams killed prey

Right The Mauritius Kestrel, the world's rarest raptor, reduced to nine at most; one of a captive pair kept in a last-ditch effort to ensure survival.

Below Spanish Imperial Eagle at the nest; only about 100 of this race are left, but recently conservationists have improved their breeding success.

Far right Two young European Sea Eagles in their nest. The northern and Norwegian populations of this species are still healthy, but the Baltic-Swedish populations are drastically reduced by pesticides and indirect persecution.

averaging 397 and 522 grams. Both male and female Goshawks could kill birds up to 1500 grams but in Europe female Goshawks kill still heavier mammals than males.

Similarly, in Europe, Goshawks and Sparrowhawks between them kill birds ranging in size from Goldcrests to cock Pheasants. The range of accipiter sizes is not so complete as in America but the female Sparrowhawk is larger than the female Sharp-shin and to some extent replaces Cooper's Hawk. The range of prey overlaps but 77·4 per cent of Sparrowhawk prey weighs less than 40 grams and 94·8 per cent of Goshawk prey more than 40 grams; both take about the same proportion of prey weighing 40 to 80 grams.

In North America and Europe, partial or complete migration of accipiters further complicates the situation, but in tropical west African forests, four species are permanently resident. They are the large Black Sparrowhawk, the medium-sized African Goshawk, the small Chestnut-bellied Sparrowhawk, and the tiny Red-thighed Sparrowhawk. Due to sexual size dimorphism, they form eight size groups, ranging from male Red-thighed Sparrowhawks weighing 78 to 82 grams to female Black Sparrowhawks weighing 650 to 790 grams. The smaller species overlap in size but the two largest do not. The more powerful foot of the African Goshawk, however, suggests that it could kill larger prey in relation to its weight, so overlapping in function with the Black Sparrowhawk.

In this case, sufficient prey could not be recorded to make the same sort of comparison possible in Europe and North America, but the hawks were trained by austringers and flown in Africa and Europe at live prey. They behaved as predicted. Black Sparrowhawks could kill anything up to Pheasant size, but would not pursue anything much smaller than a Jay. African Goshawks could kill prey of 90 to 600 grams, Chestnut-bellied Sparrowhawks 30 to 300 grams, and the tiny Red-thighed only very small birds of 8 to 40 grams. The only surprise was that while a male Black Sparrowhawk would not pursue mammals, a female did and killed a rabbit of 2500 grams and a cat. She might have been unfamiliar with such mammals in tropical forest though I doubt this. In this case, however, conjecture is being carried too far, because my experience is that Black Sparrowhawks are entirely bird eaters, in Kenya seldom killing birds weighing more than 120 grams.

In New Guinea, where eight species of accipiters occur, and in Celebes where four are found in a small area, ecological separation confines some to mountain forests and others to lowland. Moreover, examination of their bills and feet suggests that some are reptile eaters while others eat birds, so that competition in the same environment would be avoided by prey preference. No adequate details of prey taken are available, however, to prove the theory.

In harriers, the sexes are not only of different sizes but males are very differently coloured, often being grey when females are brown. In most parts of the world, only one or at most two species of harrier occur so that they cannot compete. In Europe, however, three species often breed quite close together and their possible competition has been surveyed in great detail by W J A Schipper. All three are basically small mammal eaters, varied with birds, frogs, lizards, and insects. Of the three, Montagu's Harrier is a complete migrant and the Marsh and Hen Harriers mainly resident in the areas studied. All three tend to hunt in much the same way over rank vegetation and there is much overlap in the type of prey taken. Potential competition between them should be more acute than in the three accipiters of American forests.

The harriers tended to avoid direct competition first by breeding at different times, however, the Marsh Harrier laying earliest and Montagu's latest. In breeding harriers, the female starts to hunt when the young are part grown, and thereafter brings most prey to the brood, meaning that male Montagu's and female Marsh Harriers were hunting at the same time. These were the smallest and largest of the possible size range, and because the size of prey they could kill is in proportion to body size, they did not directly compete. Competition was also avoided by different hunting methods and areas. The Hen Harrier was the most agile of the three and killed most birds in the open. Montagu's Harrier might hunt up to 12 kilometres from the nest, the Marsh Harrier usually within 2 kilometres, and females tended to hunt near the nests, males further away, over rather different vegetation. The combination of different sizes, different breeding dates, prey preferences, and the role of the sexes at different stages of the nesting cycle reduced potential competition. Harriers of different body weight and killing ability tended to be hunting together while those of similar body weight and prey preference hunted at different times and in different areas.

These examples all show how different raptors may avoid competition. In some cases, where direct competition occurs predators may exclude each other. In southern France, Bonelli's and the Golden Eagles are said to be mutually exclusive because they hunt similar prey (mainly rabbits and gamebirds) and the smaller, more active Bonelli's is able to exclude the larger Golden Eagle. Optimum habitat for Bonelli's Eagle is probably marginal for the Golden Eagle, however, and in any case, the situation in France has been so distorted by persecution that no

Above The Prairie Falcon at the nest; populations of this species have been almost doubled in Canada by construction of artificial nest cavities in gorges.

Right The Peregrine Falcon, the magnificent species which alerted the human race to the danger of organo-chlorine insecticides by its sudden decline in the 1950s; it has become extinct in eastern North America.

A young Bushbuck; about as big an animal as ever is killed by any eagle. Crowned Eagles kill them occasionally up to 20 kilograms in weight.

Wounds in my back proclaim that a female Crowned Eagle strikes with her foot wide open; fortunately she took mostly shirt.

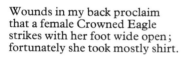

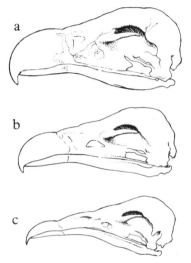

sound conclusions could be drawn. In African savannas, a rather similar, potentially competitive situation occurs between the Martial Eagle and the African Hawk-eagle, one twice the size of the other, but both gamebird eaters with a large degree of overlap. They certainly are not mutually exclusive, perhaps because they each maintain large enough ranges for their prey to be superabundant.

Competition for favoured prey may sometimes appear more acute. In the Tsavo Park in Kenya, Martial Eagles, African Hawk-eagles, Tawny Eagles, and Bateleurs all occur in the same area and all prey heavily on dik-dik, small antelopes weighing when adult 4 to 5 kilograms. Dik-dik were 40 to 47 per cent by weight of the diet of all species and the greatest overlap was between the Bateleur and the Tawny Eagle (78 per cent). C Smeenck concluded that all the breeding eagles killed 1017 to 1485 dik-dik in 110 square kilometres of the Tsavo Park. All these raptors, not to mention Leopards, Cheetahs, Serval Cats, and jackals might eat dik-dik and this suggests that dik-dik were superabundant. In September, however, a time of maximum dik-dik density, they occurred at 25.5 per square kilometre so that in the 110 square kilometres there would only have been 2800 altogether. In other words, 36 to 56 per cent of all dik-dik were eaten by the eagles, and at that rate they would soon become extinct. This result is based on only one count of dik-dik, however, so that the figures are probably unreliable. If not, the dik-dik could hardly survive as they do because they cannot multiply like voles or lemmings, but can only produce a maximum of two calves a year. It seems most likely that the dik-dik were, in fact, superabundant, so that all the raptors could prey upon them without directly competing.

Vultures are non-predatory raptors that compete for shares of the same dead animal. No detailed studies of New World vultures are available, but six species occurring in east Africa have been studied by H Kruuk. They form three species pairs, each with different functions at the carcass. The small Egyptian and Hooded Vultures have thin bills and light skulls; they are suited to picking up scraps or stripping shreds of flesh from narrow spaces which the

The searching vulture does not require marvellous eyesight to find a large carcass; European Griffon in flight.

a

b

c

Fig. 47 Skulls of African vultures (after H Kruuk):
(a) Lappet-faced Vulture – very heavy bill, heavy-boned skull, suited to tearing skin and sinew;
(b) Rüppell's Griffon – more slender skull, lighter bones, but heavy beak suited to tearing soft flesh in quantity;
(c) Egyptian Vulture – slender beak and light-boned skull, suitable for probing cavities and picking up scraps.

heavier bills of bigger vultures cannot reach. The White-backed and Rüppell's Griffons are large or very large with heavy bills and long necks, but relatively light skulls. They are anatomically suited to thrusting their long necks into bodies and tearing off soft flesh. The Lappet-faced and White-headed Vultures are large or very large, with very heavy bills and heavy skulls, adapted to tearing tough flesh off bones and eating skin. Lappet-faced Vultures can also open a carcass when no others can.

Probably none of these vultures is strongly territorial in breeding quarters and, in any case, vultures are then ecologically separated. The Egyptian is a hole nester in cliffs, the Hooded a solitary nester in trees. The White-backed is a colonial tree nester, Rüppell's Griffon a highly colonial cliff nester. The Lappet-faced and White-headed are both solitary nesters in trees but the Lappet-faced is more numerous. Usually, the White-headed is the first to arrive at a carcass; though forming only 3 per

The Hooded Vulture's slender bill can be used to extract the larva from a dung-beetle's ball.

cent of the population, it arrives first on about half the occasions, suggesting it is the chief searcher. They are followed by the Lappet-faced, then White-backed, then Rüppell's Griffons, and in the end Egyptian and Hooded Vultures remained picking up scraps, the former from bones, the Hooded more often from the ground.

Although bill structure, time of arrival at the carcass, and different habits helped to reduce competition, in fact, all six compete actively at the same carcass for a share. The Lappet-faced was more aggressive than any other, followed by Rüppell's Griffon. The specialized threat displays of large vultures were used and an aggressor always won. Similar species attacked one another more often than they attacked others and larger species attacked smaller but not vice versa. In the squabbling mass, however, each could get a share by the exercise of differing functions. The White-backed and Rüppell's Griffons were the most numerous, feeding on the soft flesh, the major, edible part

In possession, a White-backed Vulture strikes a threat posture signifying dominance.

of any large carcass for vultures. White-headed and Lappet-faced Vultures were more likely to tear at bits of skin but the larger, more powerful Lappet-faced often fed on the carcass, too, while the smaller White-headed picked up dropped pieces of flesh lying around. Thus, although there is violent, direct competition among vultures at any carcass, it is to some extent minimized by differences of structure and function. The six species formed three pairs with the functions of each member of a pair slightly different to that of the other.

All these studies of particular groups of raptors, whether temperate or tropical, tend to show that competition for prey is avoided by specialization in prey preference, varying size, different hunting methods, or structural adaptations. If such adaptations are needed to avoid competition for available prey, then it would seem that the amount of prey available must be the main factor controlling the ecology of predation. The alternative would be that certain types of prey are superabundant, so that a variety of predators can compete for a share of it without excluding one another. This, in effect, is what happens at a carcass with vultures; there is a locally superabundant mass of suitable food and all must compete actively for it while it is available. Sometimes so much is available that all the vultures in the area cannot eat it, and it rots. The differing functions have presumably evolved to assist survival in times of shortage.

Examination of individual species or individuals within species sometimes complicates the issue still further. In Scottish Golden Eagles, the density of the population varies little, whether in the heathery east, the green, grassy hills of west Argyll, or the bleak bogs of north-west Sutherland. In all, the average home range of a pair is about 4600 hectares. The amount of natural food available in these areas varies enormously, however. There are thirty or forty times as many grouse and Mountain Hares in some eastern areas as in western areas. The Eagles seem to survive as well in both, even breeding success is not much better in the east. This is sometimes explained by saying that in the west there is much more carrion available than in the east so that the shortage of live prey is made up from that. In both areas the young are reared mainly on live prey.

At the two extremes, in the same species, and in the same country, not more than 80 kilometres apart in a straight line, we could find breeding ranges supporting so many Hares and grouse that the Eagles cannot seriously limit their numbers; and others where their predation effect on a small grouse and Hare population could be limiting if these prey are always taken in the same proportion (which is improbable). In such cases, social behaviour,

Lappet-faced Vulture launching an attack, wings spread, head down, and feet stretched out; a second prepares to follow.

The rule is, the attacker always wins; here an adult Rüppell's Griffon drives away a young White-backed Vulture.

This Fish Eagle has caught a fish in thick water cabbage *(Pistia)*, and struggles to subdue it, sinking in slowly.

limiting the population of raptors to a certain number, irrespective of the total available food supply may be more important than total food supply. This seems to apply also with resident tropical eagles in African savannas; their ranges support far more potential prey than they could possibly eat. In their case, however, other predators are very numerous and the situation has not been fully studied or explained.

The behaviour of individuals of the same species in the same area complicates the issue still further. At Lake Naivasha where Fish Eagles only hunt for about one hour in a hundred, it is then incredible that total availability of food can be the limiting factor. Yet of two pairs, nesting 180 metres apart, fishing in the same lagoon, one bred three times in eighteen months rearing four young; the other not at all. In a population of Common Buzzards or Golden Eagles in Scotland, some individuals are consistently more successful. One female Buzzard may lay above average clutches year after year and rear good broods, while her next-door neighbour, in apparently similar country with as many rabbits or voles, fails to lay as often and rears fewer young. Such variations are partly, I suspect, due to the individual hunting skill of the raptor concerned and are only evened out by studying a large enough group of pairs.

Any answer to the apparently straightforward question of whether or not diurnal birds of prey control the numbers of their prey, would have to be hedged with qualifications. At times, as in the Craigheads' Michigan study area in winter, they clearly seem to do so. At other times and in other places they equally do not. At times, a species such as the Rough-legged Buzzard in the tundra, or probably Letter-winged Kites in the Australian outback, displays opportunism in the presence of abundant food, concentrates in an area, lays large clutches, and rears big broods. They make greater inroads on the available prey but even then no diurnal predator can possibly multiply in the manner of voles, lemmings, rabbits, or even songbirds which may breed several times a year. At other times, a stable and permanently resident raptor population, such as is found in African savannas, clearly cannot seriously affect the abundance of potential prey in an average home range.

In most cases, we come back to the fact that in no known case has all the arithmetic been done correctly. If the predator has been very fully studied, its prey probably

Male Sparrowhawk on his regular plucking post; we can determine what such raptors eat by examining the remains found at such places.

has not. Estimation, sometimes very rough, has to be used at some stage of the study. We can predict from a graph of food consumption against bodyweight how much a raptor should eat but this must be proved by maintaining captive birds. We can examine feet, bills, wings, tails, and even intestines (snail eaters have long intestines) to guess the type of prey eaten by a raptor and how it might hunt but we must then watch it to see if it does hunt in that way and eat what it should. Then, we must find out how many prey there are, and how they expose themselves to predation.

It would be admitted by anyone interested in this subject that the ecology of predation, whether by birds of prey or other predators, is still very imperfectly understood. Some situations, usually in cold climates, with relatively few species, seem comparatively simple. Others, for instance, the interior of the tropical forest, where the nests of many species of raptors have never yet been found, are obviously so extremely difficult that they may never be fully understood. It is best to start with a simple case and work that out thoroughly first. It should not be beyond the wit of man to find out how to count the hyrax prey of Verreaux's Eagles, or jackrabbits on the plains of Utah.

Predators and their prey in nature arrive at a balance, fluctuating this way and that, in which if the predator becomes too numerous it may eat too many of its prey, and then itself either die or be unable to breed so that the balance swings back the other way. Social behaviour dispersing resident species – the majority in all warm climates – evenly over the terrain, is evidently a method by which the numbers of predators are rather strictly limited. There are still the same three pairs of Martial Eagles in 380 square kilometres of Embu district now as there were thirty years ago, though other conditions have changed. In permanently resident species, home ranges must always support the necessary minimum number of prey for survival but the predators may not increase above a certain limit, though the prey may be far more numerous than they apparently need.

In the last analysis, also, every prey animal must not only survive predation but also disease and hunger which are frequently far more catastrophic killers than predators. The numbers of Lions in the Nairobi National Park do not vary much but in a series of good rain years the numbers of their prey build up and then in a drought the whole area is covered with stinking dead bodies, far beyond the capacity of the available hyenas and vultures to eat. The way in which such checks and balances work varies infinitely in different species of predator and prey, which explains why no-one yet fully understands the whole complex problem.

Conservation and protection

'Conservation', I am told, is no longer a fashionable word, indeed, that its use may detract from the aim of saving species from extinction because some people choose to equate it with sentimentalists or extremists, who overstate the case and do active harm by giving the enemies of conservation useful ammunition. It is necessary, therefore, to define conservation clearly. In my dictionary to conserve means *to keep from harm, decay, or loss*. A fuller biological definition is *the maintenance of the optimum flow of energy intake from the sun, and its outgoing through the growth and activity of plants and animals*. This is too wide for birds of prey alone and the simpler definition, that the conservation of birds of prey means to protect them from decay, harm, or loss is better.

It is only man who creates the need to conserve species in this sense, and only man can devise and take the necessary conservation measures. The natural dynamic balance between the incident energy of the sun, rainfall, and resultant plant growth, herbivores that eat the plants, and the predators that eat the herbivores and each other is maintained over very long periods. It is only affected by major geological catastrophes such as volcanic eruptions or earthquakes or by very gradual climatic changes which may permit some animals to adjust while others, less adaptable, become extinct. Conceivably, the California Condor was already on its way out before man pushed it toward extinction. I cannot name one raptor threatened with extinction by anything other than man and his activities. Some may locally be affected by, for instance, overpopulation of elephants which destroy big trees but such effects are indirectly due to man.

To assess the conservation action needed, we must first identify the species threatened, and the level of threat. Secondly, we must identify the cause of the threat but which may be direct, such as active persecution, or indirect, such as destruction of habitat or pesticides. Thirdly, we must estimate what should be done to combat these threats and assess its practicability. We can then see the magnitude of the problem, where the priorities lie, and how to try to set about it. Unfortunately, present-day politics and the insane belief of most of the human race that it can go on multiplying indefinitely unchecked, are likely to defeat otherwise sound conservation measures.

In the first instance, we must distinguish between species and races. There may be local geographical races which are threatened though the species as a whole is not, or is less acutely threatened. For instance, the Spanish Imperial Eagle, *Aquila heliaca adalberti*, and the Everglade Kite, *Rostrhamus sociabilis plumbeus*, are both reduced to a few hundred individuals or less. The Imperial Eagle is moderately threatened throughout its range but acutely threatened only in Spain. The Everglade or Snail Kite may be acutely threatened in Florida but is abundant elsewhere in its range, and if the Florida race is rejected by systematic revision (as it may be) the whole situation technically changes. The remaining Florida Everglade Kites obviously should still be conserved but their importance to the species as a whole is reduced.

Again, we must distinguish between mere rarity and being threatened. The Taita Falcon, *Falco fasciinucha*, is, and apparently always has been, very rare indeed for obscure reasons throughout its wide African range. The scarce population is apparently not in any way threatened, however. The South African Black Harrier, *Circus maurus*, however, is apparently rather rare in a rather restricted range and though the population seems at present stable it evidently could be threatened by drainage or development of its breeding swamps. Many species believed to be rare may on further investigation prove to be not so rare, just little known. The African race of the Lammergeier, *Gypaetus barbatus meridionalis*, was believed threatened and is in the *Red Data Book* because it is rare in south and east Africa (about 120 pairs perhaps) but it is abundant in Ethiopia where there may be 16 000 or more.

With these provisos, going through the list, I can find only four acutely threatened species out of 287, one of which may even have become extinct unnoticed. The California Condor (about forty or fifty alive) is sometimes regarded as a living fossil on its way out anyway, but its drift towards extinction has been hastened by man. The Madagascar Serpent Eagle (numbers unknown) has not been seen for many years, is apparently rare in a restricted forest habitat, and may already be extinct. It is so little known, however, that it may reappear and be found more numerous than was thought. The Philippine Monkey-eating Eagle, the second largest eagle in the world, reduced to less than 100 in the wild state, is acutely threatened and the subject of a special protection programme. Finally, most endangered of all so far as known, the Mauritius Kestrel is reduced to less than ten wild individuals, and a last-ditch effort is being made to save it by captive breeding.

Of these acutely threatened species, three are island species confined to decreasing areas of threatened forest habitat; the California Condor lives in a sort of ecological island that is near Los Angeles and San Diego.

In all, the status needs careful, regular investigation and monitoring, and they deserve first priority in efforts to save them. In the Madagascar Serpent Eagle, nothing useful is being done to investigate its status so as to start needed conservation measures. The status of the other three is quite well known and special conservation measures have been instituted. In the Philippine Monkey-eating Eagle, I doubt if the measures attempted are very effective, while in the California Condor we must reserve opinion on the success of the best-designed conservation measures because of the human pressures that beset it. If the Mauritius Kestrel can be bred in captivity, it will be saved but it can never again be as numerous as formerly because most of its habitat is already destroyed and that which remains is severely affected by human pressure and Rhesus Monkeys, again unwisely introduced by humans.

Another twenty species are less acutely threatened but also to some extent endangered. They may number hundreds or thousands but they are known to be under threat of various sorts. They include the Andean Condor, said to be decreasing, the Bald Eagle, European Sea Eagle, and Steller's Sea Eagle, all said to be, or actually decreasing through human pressures and pesticides, or both. The Lammergeier is decreasing in some of its haunts and stable in others but is always in danger because of its specialized, bone-eating habits. The European Griffon, Cape Vulture, and Black Vulture are threatened by advancing civilization, better sanitation, and better stock-keeping, reducing available carrion. The South African Black Harrier is said to be stable but needs more careful survey and study and Montagu's Harrier has decreased sharply in northern Europe since 1970. Two accipiters, the New Britain Grey-headed Goshawk and Gundlach's Hawk are both confined to possibly threatened island habitats. The absurdly confiding Galápagos Hawk found only on some of these islands, is still persecuted. The Lesser and Greater Spotted and Imperial Eagles have all decreased in parts of their range and are more or less threatened. The Seychelles Kestrel is not immediately threatened but its extremely limited habitat could obviously be endangered by any catastrophe. Three large falcons, the Saker, Gyrfalcon, and Peregrine are all endangered by the irresponsible demands of falconers or by pesticides. Obviously, some of these are worse threatened than others, and better information might show that some are not really threatened at all. It is best to be on the safe side, however.

Then, there are forty-four very little-known species which might be threatened, if the facts were known. Most are island species most of which also inhabit forest. A few are open-country species with very limited range, such as

The Galápagos Hawk; a species of *Buteo* so tame that such pictures can be taken without a hide; reduced by human persecution to about 130 pairs, but now less threatened than it was.

the Spot-winged Falconet, and a few are large and spectacular, wide-ranging species with possibly very low population density, such as the Harpy Eagle. There is no good reason to suppose that some of these are actually threatened at present but they seem next in the list of species likely to come under threat from human activity. Human population increase, now inevitable unless catastrophic famines or disease stop it, will mean that hitherto unspoiled forests in Amazonia and other places will be cultivated, and the specialized raptors living in them will die out.

You may be surprised to see that the Osprey, Golden Eagle, Marsh Harrier, and others are not in this list because in Britain and parts of Europe they may be rare, or threatened, or both. I might have included Marsh and Pallid Harriers but their breeding ranges are large and their population dynamics differ from those of large eagles. Apart from some Mediterranean countries, most Europeans and North Americans are now more aware of the need to conserve birds of prey. They are usually protected by law and active efforts are sometimes made to protect them. There may be only about ten pairs of

Ospreys in Britain but that is ten times as many as in 1955 and there are about 2500 pairs in Sweden. Golden Eagles are slowly decreasing through direct persecution and human interference in Scotland but, in their enormous range, they are probably the most numerous large eagle in the world.

In any particular country, conservationists can look at the list of raptors and see which are threatened and what should be done to conserve them. I am here trying to examine conservation needs of raptors on a global scale.

The threats to any species can include a number of factors.

1 Direct persecution – shooting, trapping, poisoning, and so on. The worst affected species are large eagles, large and small accipiters, and some falcons.

2 Indirect persecution which includes, for instance, rock climbers, campers, or fishermen who unnecessarily invade breeding haunts in spring, bird photographers who are often far too careless, and just casual walkers. These people may mean no harm but do cause harm.

3 Pesticides. Much publicized in recent times, these have initiated many studies of population dynamics. Pesticides tend to get all or most of the blame for raptor decline, whereas in Britain anyway, other human interference, direct or indirect, may do almost as much damage. Only certain organochlorine compounds have any important effect and their effect does not, in any known case, involve the whole world range of any species though it may involve the whole range of races. Moreover, not all birds of prey are equally susceptible, the most susceptible species being bird-eating accipiters and falcons, and fish-eating Osprey, sea and fish eagles.

4 Destruction of habitat, especially tropical forest. As described above, forest-loving species are usually confined to forests and specialized for life within them so that if forests are destroyed, the raptor becomes extinct. Forest habitats are severely under threat almost everywhere; they are, in fact, often looked upon as the areas which will in the future provide the surplus food needed for mankind. This hope is illusory, because the high rainfall that creates forest rapidly leaches away soil nutrients and erodes exposed soil so that the high initial fertility of forest lands is often ephemeral. The human race, however, in many forested countries is doubling every generation, and often the only as yet uninhabited areas they can invade are forests. Destruction of forests, some of which is inevitable, is probably the worst long-term threat to birds of prey. It may even be caused by foresters themselves, some of whom, prefer to see nice, regular stands of pines or eucalyptus to the jumble of beautiful and interesting, but possibly uneconomic plants that make up a tropical forest.

The various threats affecting the four most threatened species are analysed below.

1 California Condor. Direct persecution, the main cause of the original decline, is now negligible. Indirect persecution is considerable, the habitat is invaded by numbers of people, including ornithologists, and demands to make dams, recreation areas, and so on persist. Pesticides have little effect, and the main breeding habitat is conserved as a sanctuary which is, however, not an inviolable National Park. This species can only survive with great difficulty despite active conservation efforts.

2 Madagascar Serpent Eagle. The habitat is little known and breeding requirements unknown. Direct or indirect persecution and pesticides have little or no effect; probably pesticides could not affect this species because of its food – snakes. It inhabits rainforest, however, most of which, in Madagascar, has already been destroyed while what remains would be under threat. If this species still exists (it has not been seen for thirty years) and if enough forest habitat can be conserved, a small population should survive.

3 Philippine Monkey-eating Eagle. This bird was threatened by direct persecution (shooting for trophies) and collection for zoos and museums. This has now been made illegal but laws are not necessarily effective. The forest habitat has shrunk rapidly in the face of continuing exploding human population and will probably be further reduced. The species cannot regain much of its former range and can only survive in small pockets of forest if strenuous efforts are made to protect these and it. It probably has not been adequately studied in the wild state to assess its basic biology, spatial requirements, and capacity for unaided survival, and attempts to prevent forest destruction are neither adequate nor effective. This magnificent eagle will probably decrease further despite conservation measures so far taken.

4 Mauritius Kestrel. Reduced to less than twenty ten years ago and now less than ten of which two are captives. Formerly common and widespread in Mauritius, it has been reduced by forest destruction and direct persecution to a small remnant in one area. This remnant breeds poorly partly because of introduced Rhesus Monkeys and partly because it is still persecuted by Mauritians. Active conservation measures are in progress, including a last-ditch attempt to breed them in captivity, but it is touch and go whether this species will survive.

The chances of survival for any threatened species can be similarly assessed. In Appendix V I have assessed the main threats to each of the moderately threatened or possibly threatened species, and their conservation depends on countering these threats. Obviously, the threats listed earlier all can be countered if the people who create them will try. If they will, they can; it is as simple as that!

1 Direct persecution must be made illegal and the law must be enforced! At present, in many countries, the law is either too feeble, not enforced, or both. Direct persecution is normally a result of sheep rearing or game preserving, though admittedly some Cretans eat Eleonora's Falcons, and in west Africa, birds of prey are sometimes shot for food. In Britain, all birds of prey are protected but the law is not always effectively enforced. Penalties are insufficient, sometimes derisory compared to those inflicted for similar offences in, for instance, Sweden. If an American professor can be fined $3000 for illegally importing egg-whites, I want to see a Scottish landowner fined £1000 for killing a Golden Eagle.

In this connection, game preservers and stock raisers, notably sheep men, will inevitably complain that raptors damage their interests. Many British birds of prey are or were persecuted because they ate homing pigeons (Peregrine Falcon), sheep and lambs (Golden and Sea Eagles –

Direct persecution; the infamous pole trap, outlawed in 1904 because of its vile cruelty, but still in use on many British estates.

Eagles are frequently accused of killing lambs, usually without reason; in this case the bruising and bleeding shows that the lamb was alive when seized by the eagle.

the latter rendered extinct for this reason), or gamebirds (amost every species – several rendered extinct or nearly so for this reason).

In virtually every case where they have been fully investigated, these complaints have been found groundless or at best greatly exaggerated. Peregrine Falcons were proved to take only about one in 800 domestic pigeons. The Golden Eagle, shot in thousands from aeroplanes in the United States, eats hardly any lambs there, but feeds mainly (98 per cent) on jackrabbits, direct competitors for sheep fodder. The habit of picking up available carrion leads to many unfounded allegations of lamb killing. It is easy to prove if a lamb has been killed by an eagle; if it has, it will have bruised and bleeding talon wounds. Sparrowhawks scarcely ever eat Pheasants (which are too big), and any damage they do to Pheasant poults in release pens can easily be avoided by siting the pens some distance from the regularly used nesting sites.

Conservationists can defeat their own aims by claiming that birds of prey never do any damage. Some do unquestionably eat grouse or Pheasants, or even domestic poultry (commonly in tropical countries). It may be necessary to admit frankly that some damage occurs and make constructive suggestions to prevent it. Often it can be prevented quite simply, at other times the inevitability of some damage may have to be accepted. Conservationists will then do well to admit that it occurs and thank those prepared to accept some such losses for the common good.

2 Indirect persecution. This is more difficult to deal

with because it is seldom deliberate. Rock climbers do not necessarily mean to make an eagle or Peregrine or Lammergeier desert, and fishermen do not necessarily mean to prevent a Sea Eagle or Osprey laying eggs. Bird photographers, however, normally know quite well that they may be exposing a rare species to unnecessary risk in taking their pictures. Like direct persecution, indirect persecution principally occurs in developed countries, and is worsened by the affluent society, with its roads, cars, speedboats, and aeroplanes, enabling much easier access to areas previously undisturbed. The swarms of walkers now penetrating to the Cairngorm nature reserve in Scotland have resulted in poor to no breeding in several formerly regular pairs of Golden Eagles. They mean no harm but the Eagles just cannot live with them.

We cannot entirely prevent such effects but perhaps damage could be reduced by restricting public access to certain areas. Any possible advantage of this would have to be weighed against the risk of deliberately revealing where a rare raptor was nesting. I think that the people who might deliberately persecute such a bird, the pigeon fanciers, falconers, gamekeepers, and egg collectors, know of such places anyway so that publicizing them could do little harm. It is at least worth a try in few places where raptors now regularly fail to breed because of inadvertent human interference.

Immature Goshawk on the fist; taking such relatively common species for falconry presents no threat, and such birds may have escaped to found a new population in Britain in recent years.

Indirect persecution includes changes in land use; in Britain notably afforestation of formerly open moorlands, which affects or may affect Golden Eagles, Red Kites, and Merlins by depriving them of normal hunting habitat. Here we must weigh the undoubted advantages against some disadvantages. Obviously, the effect of forestry is likely to be much more serious if it retards the painfully slow recovery of the Welsh Red Kite from near extinction than if it stops a marginal pair or two of Golden Eagles from breeding. Fair exchange, in the form of more breeding Hen Harriers and Short-eared Owls, even perhaps Goshawks, would in this case, be no robbery. Afforestation, broadly speaking, enriches the environment, and it is not entirely the Forestry Commission's fault that they grow square blocks of Alaskan Spruce trees for clear felling and seldom produce a mature forest. The public, who pay for it all, should demand, and see that they get, the latter. They could.

3 Pesticides. In general, this threat is again worst in developed countries though there is the very real threat that, as the more toxic pesticides are phased out in these, they may be dumped on underdeveloped countries that may be quite willing to buy them as a cheap solution to agricultural pests. I can remember when DDT was hailed with relief as a superior, relatively non-toxic insecticide, far less damaging than arsenic in its direct effects on wild life; no-one then foresaw its long-term effects. I have also, for many years, been a practising agriculturist dealing with small peasant farmers, so I can see both sides of this question. My opinion is that insecticide manufacturers have sometimes been unjustifiably condemned.

Nevertheless, it is now generally accepted as proven by most reasonable people that certain organochlorine pesticides cause certain raptors, notably Ospreys, sea eagles, falcons, and accipiters, to lay thin-shelled, easily broken eggs so that the normal reproductive biology of the species collapses and it becomes locally extinct. This has happened to the Peregrine in the eastern United States and in southern Britain, and to the European Sea Eagle, Osprey, Bald Eagle, and other fish-eating birds in parts of their range. I know of no species threatened with total extinction through pesticides. Even many of those most affected, such as the Sparrowhawk, Peregrine Falcon, and European Sea Eagle are still quite common and have healthy unaffected populations in other parts of their world range. The whole blame for recent declines is often rather unjustly laid on pesticides though it is at least aggravated by other human disturbance and by continuing direct persecution.

The more toxic pesticides are being withdrawn from the developed countries; DDT is outlawed in many. World food needs dictate that they must be replaced by practical and cheap alternatives. In time, rather a long time, these old organochlorine pesticides will disappear from the environment and, if the alternatives are not in some unforeseen manner worse, Peregrines may again breed in the Adirondacks and on the cliffs of Dover.

4 Forest destruction. There are some 4000 million people in the world of whom 2800 million live in underdeveloped countries. They are increasing at 2·5 to 3 per cent per annum, doubling in about twenty-five to thirty years. They are mainly subsistence farmers, needing about a fifth of a hectare per head to survive at prevailing low yields. Anyone can do the arithmetic on the back of an envelope but it amounts to the fact that, annually, about fourteen million new hectares should come under cultivation to maintain even the present low standard of living for peasant agriculturists.

In the past, inadequate attempts were often made to conserve necessary mountain forests as timber reserves, as protection for watersheds, even nature reserves. Hungry people are increasingly likely to ignore these reserves. Richer habitats, especially forests, will inevitably come under increasing pressure in the next two decades and there is as yet no real sign that the need for a sound, biologically based, conservative land use is understood.

Of course, not only raptors are concerned. Many other forest birds and animals, not to mention trees and plants themselves, are affected. A breeding pair of raptors, especially the larger species, however, requires a certain area to maintain them, and in many forest species this is unknown because their nests have never been found. All we can do in these cases is to study the most vulnerable species and try to judge how much forest should be conserved in order to maintain viable populations of, say, the Celebes Crested Goshawk, or Wallace's Hawk-eagle. Conservationists must then think of additional reasons for conserving such tracts of forest because they will not be conserved just for raptors. Fortunately, this is not difficult; remaining forests are often on steep mountainsides, which ought to be protected from cultivation anyway. They can produce timber and many other products, and in doing so, conserve their native inhabitants including raptors.

Unfortunately, agriculturists look at natural forests as prime agricultural land wasted by a mass of nasty timber growing on it, and foresters themselves may look at them as areas where prime crops of building poles or pit props or paper pulp could be produced. They may even think it necessary to grow such crops to protect the forests from the more rapacious agriculturists. Again, we can only hope that common sense will in the end prevail and adequate areas of natural forest will be conserved. Meantime, the outlook for raptors or anything else, including human

Birds of prey can be helped, in some cases, to recolonize or colonize new ground; a Kestrel nest box in otherwise unsuitable breeding terrain.

beings dependent on rather small areas of natural forest, is bleak.

The measures needed to combat the threats faced by raptors over the next twenty years (by which time human population increase should be settled one way or another) can be summed up as:

1 improve the laws, and penalize offenders more vigorously;
2 educate the public;
3 withdraw all pesticides that are, or might be threats to the environment;
4 conserve adequate tracts of forests and other unusual habitats, notably marshes, for the good of mankind as well as for raptors.

These are all quite common-sense and obvious measures but they often receive scant attention or at best lip service.

Some positive steps can be taken to help threatened species. The first is the provision of new nesting sites. The Emperor Menelik, when he commanded the Ethiopian peasantry to grow introduced eucalypts all over the barren, grassy plateau, inadvertently provided nesting places for crows, Augur Buzzards, Lanner Falcons, Tawny Eagles, and others, and similar effects have been noted in the United States, on the great plains of Colorado and Montana. Kestrels will readily use nesting boxes, and a classic study of the species has been done by A J Cave in Holland entirely in artificial nesting boxes. With no possible breeding place on a polder, the population rose to average densities of thirty-three pairs on 18 square kilometres, higher in some years than others. Near Hawk Mountain in Pennsylvania, many nest boxes are occupied by Kestrels. Richard Fyfe, in Saskatchewan, by making nesting holes in cliffs for Canada Geese, more than doubled the local population of Prairie Falcons. Ospreys in Wisconsin and Snail Kites in Florida have each accepted artificial metal nest structures – though in the case of the Kites they had to start their own nest first.

Quite evidently, the population of some raptors could enormously increase if helped by human beings in this way. These experiments are also of great interest in establishing the upper limit of raptor population an area can support. The Craigheads might have come to different conclusions about summer owl and Kestrel ranges and numbers if they had saturated every woodlot with suitable nest boxes.

In desperate cases, several species can now be bred in captivity. This can be tried with the desperately endangered Mauritius Kestrel because the techniques have been proved with commoner American Kestrels first. The Peregrine Falcon, extinct in the eastern United States, is being successfully bred at Cornell University by Tom Cade and his colleagues. Females can be induced to lay more than one clutch so that by incubating the first artificially, each pair effectively rears more young than they would in nature. Even artificial insemination has been tried success-

Two young Lesser Spotted Eagles about to leave the nest. Normally, the second would never have survived the Cain and Abel battle when downy; but here has been reared to near maturity by a Black Kite and returned to the nest by conservationists seeking to increase the breeding success of a decreasing species.

fully, and by selection a super-kestrel may be bred, capable of killing quail, so reducing the demand for scarcer Prairie Falcons. Such techniques, scoffed at a few years ago as impractical, are now a proven success. They are not an excuse for all to keep birds of prey, however, but should be confined to a few places where experts have adequate money and facilities. An amateur hawk-buff in his backyard has not.

Another technique, recently developed by Bernt Meyburg with the Lesser Spotted Eagle, partially overcomes apparently unnecessary losses of second-hatched young through the Cain and Abel Battle. He first took young Spotted Eagles and had them reared by Black Kites to near maturity, then replaced the big young one in the nest with its companion. If old enough, they no longer fight and the parent accepts them readily and feeds them. Evidently this process, in some eagles, can nearly double the breeding success. The technique once established, Meyburg, working in Spain with the endangered Spanish Imperial Eagle (which quite often lays three and occasionally even four eggs) transferred young with no natural chance of survival, and presented them about hatching time to females with infertile eggs. The females accepted and readily reared these young and one year's work with a few pairs increased the breeding success of these pairs by 40 per cent, and that of the total Spanish population by about 4 per cent. Evidently, if this were universally applied all over Spain it could greatly increase the productivity of Spanish Imperial Eagles. It has yet to be proved, however, that greater numbers of young would survive better to increase the population as a whole.

Evidently, also, this technique could be used to re-introduce non-migratory species locally rendered extinct. In Britain, the Osprey re-established itself from surplus members of the migrant Scandinavian breeding population. The Sea Eagle, however, extinct since 1916 in Britain, but formerly quite common round the coasts of Scotland, is unlikely to do the same because the Scandinavian population is permanently resident and is also decreasing. One attempt to re-introduce the Sea Eagle, by re-leasing four young on Fair Isle, has already failed but there seems no reason why Sea Eagle eggs or chicks, perhaps from eyries regularly lost anyway in Scandinavia, should not be placed and reared in Golden Eagle eyries instead of the young Golden Eagles – which could be hand reared.

Such an attempt would have to be done near the coast, say in Rhum, Islay, Skye, or parts of western Ross, and should be repeated in about twenty Golden Eagle eyries for at least ten years. A small but viable population of Sea Eagles might then have developed in areas they formerly inhabited, if they were not at once persecuted. Similarly, once the Welsh Red Kite has built up to greater numbers as it eventually promises to do, eggs or nestlings could be reared by Common Buzzards in parts of Argyll or Devon where the conditions for the Kites appear ideal and where they once were common. Any such positive efforts to re-introduce species which have been locally exterminated, however, would need much better support from the law enforcement bodies than they are likely to get in Britain at present.

Some species are already themselves taking advantage of man's activities, as discussed under towns previously. They may need a little help, such as the provision of suitable nesting boxes for Peregrines and other large falcons. Hen Harriers are the only species actively increasing in Britain at the moment, largely because of the increasing areas under young forestry plantations. This was not what the foresters intended but surely British forests could be managed so that there is always some suitable breeding habitat for Montagu's Harriers and Hen Harriers in the young stages and for Buzzards and Goshawks in mature forests. There are, of course, reasons that would be advanced against any such thing by game preservers against Goshawks, and by the foresters themselves on economic grounds, for mature old woods can only be produced by thinning and a longer rotation which may be technically uneconomic.

Conservation of birds of prey boils down to a few simple steps, preventive and constructive. Persecution must be effectively stopped where it continues and further education should be undertaken to ensure that the role of birds of prey in nature is understood. Pollution of the environment, undesirable anyway, should be controlled, and forests and other habitats, worldwide, managed sensibly so that their original inhabitants may not be lost. Enthusiasts can take positive steps to increase nesting opportunities by providing new nesting places, or increase breeding success by substituting young in nests. Some species can be bred in captivity and released again later. Ten years ago, some of these techniques had not been developed; now we seem only on the threshold of new and hopeful developments.

Such proposals may seem Utopian when we consider how rapidly world population is growing to what may be self-destructive levels. If we do not take a more sensible and conservation-minded view of land use, however, we are doomed to catastrophe. The ruination of the last few hectares of raptor habitat in tropical forests will not prevent this. When next you see a pole trap, or some other activity that threatens birds of prey, you may legitimately echo John Donne, 'And therefore never send to know for whom the bell tolls; it tolls for thee!'

Appendices

APPENDIX I
The living species of the order Falconiformes

This list of species is based upon the specific list in Brown and Amadon's *Eagles, Hawks and Falcons of the World* with a few minor changes which seem clearly desirable in the light of recent research. In the full revision of the Peters' List of Falconiformes now being prepared by Dr Amadon himself, many other changes of sequence or generic or specific nomenclature are likely to be proposed, and we do not wish to anticipate these. The systematic changes made here are minor, involving the genera *Aegypius*, *Torgos*, *Sarcogyps*, and *Trigonoceps* (all now merged in *Aegypius*), and *Lophaetus* (now included in *Spizaetus*). The Augur Buzzard, *Buteo augur*, is separated here as a full species from the Jackal Buzzard, *Buteo rufofuscus*, making 287 species in all. The final number in the revised Peters' List is likely to be more than 290, with fewer genera, in somewhat different sequence, and with species in large genera such as *Accipiter*, *Buteo*, or *Falco* also in different sequence.

To save repetition of the same names in another appendix, condensed details on the level of knowledge of breeding biology (see Chapter) have been included here. The letters following each name (A, B, C, D, E) indicate whether a species is intimately known, very well known, well known, little known, or unknown in this respect. The figures in brackets refer to details of clutch size so far as is known, with the average given to the nearest whole egg where the clutch varies, thus [1–3(2)]; a query after the number indicates that the clutch size is thought to be, say, 2 but that this is not certain.

ORDER FALCONIFORMES
SUBORDER CATHARTAE (NEW WORLD VULTURES)
SUPERFAMILY CATHARTOIDEA
Family Cathartidae

Genus *Cathartes*

Cathartes aura	Turkey Vulture	B (2)
C. burrovianus	Yellow-headed Vulture	D (2?)
C. melambrotus	Greater Yellow-headed Vulture	E

Genus *Coragyps*

Coragyps atratus	Black Vulture	C (2)

Genus *Sarcorhamphus*

Sarcorhamphus papa	King Vulture	D (1)

Genus *Gymnogyps*

Gymnogyps californianus	California Condor	B (1)

Genus *Vultur*

Vultur gryphus	Andean Condor	C (1)

SUBORDER ACCIPITRES
SUPERFAMILY ACCIPITROIDEA
Family Pandionidae (Osprey)

Genus *Pandion*

Pandion haliaetus	Osprey	A [2–4 (3)]

Family Accipitridae (kites, Old World vultures, hawks, buzzards, eagles, etc.)

(SUBFAMILY PERNINAE)

Genus *Aviceda*
Aviceda cuculoides	African Cuckoo Falcon	C (2–3)
A. madagascariensis	Madagascar Cuckoo Falcon	E
A. jerdoni	Jerdon's Baza	C (2–3)
A. subcristata	Crested Baza	C (2–3 occ. 4)
A. leuphotes	Black Baza	C (2–3)

Genus *Leptodon*
Leptodon cayanensis	Cayenne Kite	E

Genus *Chondrohierax*
Chondrohierax uncinatus	Hook-billed Kite	D (2–3)

Genus *Henicopernis*
Henicopernis longicauda	Long-tailed Honey Buzzard	E
H. infuscata	Black Honey Buzzard	E

Genus *Pernis*
Pernis apivorus	Honey Buzzard	A [1–3 (2)]
P. celebensis	Barred Honey Buzzard	E (2?)

Genus *Elanoides*
Elanoides forficatus	American Swallow-tailed Kite	B [2–4 (3)]

(SUBFAMILY MACHAERHAMPHINAE)

Genus *Machaerhamphus*
Machaerhamphus alcinus	Bat Hawk	C (1–2)

(SUBFAMILY ELANINAE)

Genus *Gampsonyx*
Gampsonyx swainsonii	Pearl Kite	D (3)

Genus *Elanus*
Elanus leucurus	White-tailed Kite	B (4–5)
E. caeruleus	Black-shouldered Kite	B (3–5)
E. notatus	Australian Black-shouldered Kite	C (3–4)
E. scriptus	Letter-winged Kite	C (3–6)

Genus *Chelictinia*
Chelictinia riocourii	African Swallow-tailed Kite	C (4)

(SUBFAMILY MILVINAE)

Genus *Rostrhamus*
Rostrhamus sociabilis	Snail Kite	C [2–4 (3)]
R. hamatus	Slender-billed Kite	D

Genus *Harpagus*
Harpagus bidentatus	Double-toothed Kite	D (1–2)
H. diodon	Rufous-thighed Kite	D (2)

Genus *Ictinia*
Ictinia plumbea	Plumbeous Kite	C (1–2)
I. misisippiensis	Mississippi Kite	B (1–3)

Genus *Lophoictinia*
Lophoictinia isura	Square-tailed Kite	C (2)

Genus *Hamirostra*
Hamirostra melanosternon	Black-breasted Buzzard Kite	C (2)

Genus *Milvus*
Milvus migrans	Black, Common, or Pariah Kite	B [1–5 (2–3)]
M. milvus	Red Kite	B [1–5 (2–3)]

Genus *Haliastur*
Haliastur sphenurus	Whistling Eagle, Whistling Hawk	C (2–3)
H. indus	Brahminy Kite, White-headed Sea Eagle	C [1–4 (2–3)]

(SUBFAMILY HALIAEETINAE)
Genus *Haliaeetus*

Haliaeetus leucogaster	White-bellied Sea Eagle	C (2–3)
H. sanfordi	Sanford's Sea Eagle	E
H. vocifer	African Fish Eagle	B [1–3 (2)]
H. vociferoides	Madagascar Fish Eagle	C [1–3 (2)]
H. leucoryphus	Pallas' Sea Eagle	C [1–3 (2)]
H. leucocephalus	Bald Eagle	B [1–3 (2)]
H. albicilla	European Sea Eagle	B [1–3 (2)]
H. pelagicus	Steller's Sea Eagle	C [1–3 (2)]

Genus *Ichthyophaga*

Ichthyophaga nana	Lesser Fishing Eagle	C [1–3 (2)]
I. ichthyaetus	Grey-headed Fishing Eagle	C [2–4 (3)]

Genus *Gypohierax*

Gypohierax angolensis	Vulturine Fish Eagle	C (1)

(SUBFAMILY AEGYPIINAE)
Genus *Neophron*

Neophron percnopterus	Egyptian Vulture	B [1–3 (2)]

Genus *Gypaetus*

Gypaetus barbatus	Lammergeier, Bearded Vulture	C (1–2 occ. 3)

Genus *Necrosyrtes*

Necrosyrtes monachus	Hooded Vulture	C (1)

Genus *Gyps*

Gyps bengalensis	Indian White-backed Vulture	C (1)
G. africanus	African White-backed Vulture	B (1)
G. indicus	Long-billed Vulture	C (1)
G. rueppellii	Rüppell's Griffon	B (1?2)
G. himalayensis	Himalayan Griffon	C (1)
G. fulvus	Griffon Vulture	B (1)
G. coprotheres	Cape Vulture	C (1)

Genus *Aegypius*

Aegypius tracheliotus	Lappet-faced Vulture	C (1)
A. calvus	Indian Black or King Vulture	C (1)
A. monachus	Black or Cinereous Vulture	B (1)
A. occipitalis	White-headed Vulture	C (1)

(SUBFAMILY CIRCAETINAE)
Genus *Circaetus*

Circaetus gallicus	Short-toed Eagle, Serpent Eagle	A (1)
C. cinereus	Brown Snake Eagle	B (1)
C. fasciolatus	Southern Banded Snake Eagle	D (1)
C. cinerascens	Smaller Banded Snake Eagle	C (1)

Genus *Terathopius*

Terathopius ecaudatus	Bateleur	B (1)

Genus *Spilornis*

Spilornis holospilus	Philippine Serpent Eagle	E
S. rufipectus	Celebes Serpent Eagle	E
S. cheela	Crested Serpent Eagle	C (1)
S. klossi	Nicobar Serpent Eagle	E
S. elgini	Andaman Serpent Eagle	E

Genus *Dryotriorchis*

Dryotriorchis spectabilis	Congo Serpent Eagle	E

Genus *Eutriorchis*

Eutriorchis astur	Madagascar Serpent Eagle	E

(SUBFAMILY POLYBOROIDINAE)
Genus *Polyboroides*

Polyboroides typus	African Harrier Hawk	B (1–2)
P. radiatus	Madagascar Harrier Hawk	C (1–2)

Genus *Geranospiza*

Geranospiza caerulescens	Crane Hawk	D (1–2?)

(SUBFAMILY CIRCINAE)
Genus *Circus*
Circus assimilis	Spotted Harrier	C [2–4 (3)]
C. aeruginosus	Marsh Harrier, Swamp Hawk	B (3–8)
C. ranivorus	African Marsh Harrier	C (3–5)
C. maurus	Black Harrier	D (3–5)
C. cyaneus	Hen Harrier, Marsh Hawk	A [3–12 (4–6)]
C. cinereus	Cinereous Harrier	D (3–4)
C. macrourus	Pale Harrier	B (3–6)
C. pygargus	Montagu's Harrier	B [3–10 (4–5)]
C. melanoleucus	Pied Harrier	B (4–6)
C. buffoni	Long-winged Harrier	D (3–5?)

(SUBFAMILY ACCIPITRINAE)
Genus *Melierax*
Melierax metabates	Dark Chanting Goshawk	C (1–2)
M. canorus	Pale Chanting Goshawk	C (1–2)
M. gabar	Gabar Goshawk	C [1–3 (2–3)]

Genus *Megatriorchis*
Megatriorchis doriae	Doria's Goshawk	E

Genus *Erythrotriorchis*
Erythrotriorchis radiatus	Red Goshawk	C [1–3 (2–3)]

Genus *Accipiter*
Accipiter gentilis	Northern Goshawk	A [1–5 (3)]
A. henstii	Henst's Goshawk	D [2 (2–3)]
A. melanoleucus	Black Sparrowhawk	B (2–3)
A. meyerianus	Meyer's Goshawk	E
A. buergersi	Buerger's Goshawk	E
A. ovampensis	Ovampo Sparrowhawk	C (2–3)
A. madagascariensis	Madagascar Sparrowhawk	D (3)
A. gularis	Japanese Lesser Sparrowhawk	C (3)
A. virgatus	Besra Sparrowhawk	C [2–5 (3–4)]
A. nanus	Celebes Little Sparrowhawk	D
A. rhodogaster	Vinous-breasted Sparrowhawk	E
A. erythrauchen	Mollucan Sparrowhawk	E
A. cirrhocephalus	Collared Sparrowhawk	C (2–4)
A. brachyurus	New Britain Sparrowhawk	E
A. nisus	European Sparrowhawk	A [2–7 (5)]
A. rufiventris	Rufous-breasted Sparrowhawk	C (2–4)
A. striatus	Sharp-shinned Hawk	B (4–5 +)
A. erythropus	Red-thighed Sparrowhawk	D (2)
A. minullus	African Little Sparrowhawk	B (2–3)
A. castanilius	Chestnut-bellied Sparrowhawk	E
A. tachiro	African Goshawk	B (2–3)
A. trivirgatus	Crested Goshawk	C (2–3)
A. griseiceps	Celebes Crested Goshawk	D
A. trinotatus	Spot-tailed Accipiter	E
A. luteoschistaceus	Blue and Grey Sparrowhawk	E
A. fasciatus	Australian Sparrowhawk	C (2–4)
A. henicogrammus	Gray's Goshawk	E
A. novaehollandiae	Variable Goshawk, White Goshawk, Grey Goshawk	C (2–4)
A. griseogularis	Grey-throated Goshawk	E
A. melanochlamys	Black-mantled Accipiter	E
A. imitator	Imitator Sparrowhawk	E
A. albogularis	Pied Goshawk	E
A. haplochrous	New Caledonia Sparrowhawk	E
A. rufitorques	Fiji Goshawk	D (1–3)
A. poliocephalus	New Guinea Grey-headed Goshawk	E
A. princeps	New Britain Grey-headed Goshawk	E
A. soloensis	Grey Frog Hawk	C [2–5 (3–4)]
A. brevipes	Levant Sparrowhawk	C (3–5)
A. badius	Shikra	C [2–7 (3–5)]
A. butleri	Nicobar Shikra	D
A. francesii	France's Sparrowhawk	C (3–4)
A. collaris	American Collared Sparrowhawk	E

A. superciliosus	Tiny Sparrowhawk	D (3)
A. gundlachi	Gundlach's Sparrowhawk	E
A. cooperii	Cooper's Hawk	A [3–6 (4–5)]
A. bicolor	Bicoloured Sparrowhawk	D (4)
A. poliogaster	Grey-bellied Goshawk	E

Genus *Urotriorchis*
Urotriorchis macrourus	African Long-tailed Hawk	E

(SUBFAMILY BUTEONINAE)
Genus *Butastur*
Butastur rufipennis	Grasshopper Buzzard-eagle	D [1–3 (2–3)]
B. liventer	Rufous-winged Buzzard-eagle	C (2–3)
B. teesa	White-eyed Buzzard	C (2–3)
B. indicus	Grey-faced Buzzard-eagle	C (2–3)

Genus *Kaupifalco*
Kaupifalco monogrammicus	Lizzard Buzzard	C [1–3 (2–3)]

Genus *Leucopternis*
Leucopternis schistacea	Slate-coloured Hawk	E
L. plumbea	Plumbeous Hawk	E
L. princeps	Prince's or Barred Hawk	E
L. melanops	Black-faced Hawk	E
L. kuhli	White-browed Hawk	E
L. lacernulata	White-necked Hawk	E
L. semiplumbea	Semiplumbeous Hawk	D
L. albicollis	White Hawk	D (1)
L. occidentalis	Grey-backed Leucopternis	D
L. polionota	Mantled Hawk	E

Genus *Buteogallus*
Buteogallus anthracinus	Common Black Hawk	D (1–2)
B. aequinoctialis	Rufous Crab Hawk	D (1–2)
B. urubitinga	Great Black Hawk	D (1)

Genus *Harpyhaliaetus*
Harpyhaliaetus solitarius	Black Solitary Eagle	D (1)
H. coronatus	Crowned Solitary Eagle	E

Genus *Heterospizias*
Heterospizias meridionalis	Savanna Hawk	C (1–2)

Genus *Busarellus*
Busarellus nigricollis	Fishing Buzzard	D

Genus *Geranoaetus*
Geranoaetus melanoleucus	Grey Eagle-buzzard	D (1–3)

Genus *Parabuteo*
Parabuteo unicinctus	Bay-winged Hawk	C (2–4)

Genus *Buteo*
Buteo nitidus	Grey Hawk, Mexican Goshawk	C [1–3 (2)]
B. magnirostris	Roadside Hawk	D (1–2)
B. leucorrhous	Rufous-thighed Hawk	D
B. ridgwayi	Ridgway's Hawk	D (1–2)
B. lineatus	Red-shouldered Hawk	C [2–5 (3–4)]
B. platypterus	Broad-winged Hawk	C (2–4)
B. brachyurus	Short-tailed Hawk	B (2–4)
B. albonotatus	Zone-tailed Hawk	C [1–3 (2)]
B. solitarius	Hawaiian Hawk	D
B. swainsonii	Swainson's Hawk	B [2–4 (3)]
B. galapagoensis	Galápagos Hawk	B (1–2)
B. albicaudatus	White-tailed Hawk	C [1–3 (2)]
B. polyosoma	Red-backed Buzzard	D (1–3)
B. poecilochrous	Gurney's Buzzard	E
B. jamaicensis	Red-tailed Hawk	B [1–4 (2–3)]
B. ventralis	Red-tailed Buzzard	D (2–3?)
B. buteo	Common Buzzard	A [1–5 (2–3)]
B. oreophilus	African Mountain Buzzard	C (2)
B. brachypterus	Madagascar Buzzard	C (2)

B. lagopus	Rough-legged Buzzard	B (2–7)
B. rufinus	Long-legged Buzzard	C [2–5 (2–3)]
B. hemilasius	Upland Buzzard	C [2–5 (2–3)]
B. regalis	Ferruginous Hawk	B (3–5)
B. auguralis	African Red-tailed Buzzard	C (2–3)
B. rufofuscus	Jackal Buzzard	C (1–3)
B. augur	Augur Buzzard	B (1–3)

Genus *Morphnus*
Morphnus guianensis	Guiana Crested Eagle	E

Genus *Harpia*
Harpia harpyja	Harpy Eagle	C (2)

Genus *Harpyopsis*
Harpyopsis novaeguineae	New Guinea Harpy Eagle	E

Genus *Pithecophaga*
Pithecophaga jefferyi	Philippine Monkey-eating Eagle	B (1)

(SUBFAMILY AQUILINAE)
Genus *Ictinaetus*
Ictinaetus malayensis	Indian Black Eagle	C (1–2)

Genus *Aquila*
Aquila pomarina	Lesser Spotted Eagle	A [1–3 (2)]
A. clanga	Greater Spotted Eagle	B [1–3 (2)]
A. rapax	Tawny Eagle, Steppe Eagle	B [1–3 (2)]
A. heliaca	Imperial Eagle	B [1–4 (2)]
A. wahlbergi	Wahlberg's Eagle	A (1)
A. gurneyi	Gurney's Eagle	E
A. chrysaetos	Golden Eagle	A [1–3 (2)]
A. audax	Wedge-tailed Eagle	B [1–3 (2)]
A. verreauxi	Verreaux's Eagle, Black Eagle	A [1–3 (2)]

Genus *Hieraaetus*
Hieraaetus fasciatus	Bonelli's Eagle, African Hawk-eagle	A [1–3 (2)]
H. pennatus	Booted Eagle	B [1–2 (2)]
H. morphnoides	Little Eagle	C [1–2 (2)]
H. dubius	Ayres' Hawk-eagle	A [1 (2?)]
H. kienerii	Chestnut-bellied Hawk-eagle	D (1)

Genus *Spizastur*
Spizastur melanoleucus	Black and White Hawk-eagle	D (1)

Genus *Spizaetus*
Spizaetus occipitalis	Long-crested Eagle	B (1–2)
S. africanus	Cassin's Hawk-eagle	D
S. cirrhatus	Changeable Hawk-eagle	C (1)
S. nipalensis	Mountain Hawk-eagle	C (1)
S. bartelsi	Java Hawk-eagle	E
S. lanceolatus	Celebes Hawk-eagle	D
S. philippensis	Philippine Hawk-eagle	E
S. alboniger	Blyth's Hawk-eagle	E
S. nanus	Wallace's Hawk-eagle	E
S. tyrannus	Black Hawk-eagle	D
S. ornatus	Ornate Hawk-eagle	D

Genus *Stephanoaetus*
Stephanoaetus coronatus	Crowned Eagle	A (1–2)

Genus *Oroaetus*
Oroaetus isidori	Isidore's Eagle	C [1 (2?)]

Genus *Polemaetus*
Polemaetus bellicosus	Martial Eagle	B (1)

SUPERFAMILY SAGITTAROIDEA
Family Sagittariidae (Secretary Bird)
Genus *Sagittarius*
Sagittarius serpentarius	Secretary Bird	B (1–3)

SUBORDER FALCONES
Family Falconidae (caracaras, milvagos, forest falcons, falconets, and falcons)
(SUBFAMILY POLYBORINAE)
Genus *Daptrius*

Daptrius ater	Yellow-throated Caracara	E
D. americanus	Red-throated Caracara	D

Genus *Phalcoboenus*

Phalcoboenus carunculatus	Carunculated Caracara	E
P. megalopterus	Mountain Caracara	C (2–3)
P. albogularis	Darwin's Caracara	D
P. australis	Forster's Caracara	C (2–3)

Genus *Polyborus*

Polyborus plancus	Common Caracara	C [2–4 (2–3)]

Genus *Milvago*

Milvago chimango	Chimango	C (2–3)
M. chimachima	Yellow-headed Caracara	D (2)

(SUBFAMILY HERPETOTHERINAE)
Genus *Herpetotheres*

Herpetotheres cachinnans	Laughing Falcon	D (1)

Genus *Micrastur*

Micrastur ruficollis	Barred Forest Falcon	E
M. plumbeus	Sclater's Forest Falcon	E
M. mirandollei	Slaty-backed Forest Falcon	E
M. semitorquatus	Collared Forest Falcon	E
M. buckleyi	Traylor's Forest Falcon	E

(SUBFAMILY FALCONINAE)
Genus *Spiziapteryx*

Spiziapteryx circumcinctus	Spot-winged Falconet	E

Genus *Poliohierax*

Poliohierax semitorquatus	African Pygmy Falcon	A (2–3)
P. insignis	Fielden's Falconet	E

Genus *Microhierax*

Microhierax caerulescens	Red-legged Falconet	C [4–5 (3–4)]
M. fringillarius	Black-legged Falconet	C [2–5 (3–4)]
M. latifrons	Bornean Falconet	E
M. erythrogonys	Philippine Falconet	D
M. melanoleucus	Pied Falconet	D (3–4)

Genus *Falco*

Falco naumanni	Lesser Kestrel	B [3–6 (4–5)]
F. rupicoloides	Greater or White-eyed Kestrel	C [2–6 (3–4)]
F. alopex	Fox Kestrel	D
F. sparverius	Sparrowhawk, American Kestrel	B [3–7 (4–5)]
F. tinnunculus	Kestrel	A [2–9 (4–6)]
F. newtoni	Madagascar Kestrel	D
F. punctatus	Mauritius Kestrel	C (2–3)
F. araea	Seychelles Kestrel	C (1–2)
F. moluccensis	Molucca Kestrel	D [4 (3–5 ?)]
F. cenchroides	Australian or Nankeen Kestrel	C (4–6)
F. ardosiaceus	Grey Kestrel	C (3–5)
F. dickinsoni	Dickinson's Kestrel	C (2–3)
F. zoniventris	Madagascar Banded Kestrel	D
F. vespertinus	Red-footed Falcon	A [2–5 (3–4)]
F. chicquera	Red-headed Falcon	C (3–5)
F. columbarius	Merlin, Pigeon Hawk	A [2–7 (5–6)]
F. berigora	Brown Hawk	C (2–4)
F. novaezeelandiae	New Zealand Falcon	C (2–3)
F. subbuteo	Hobby	B (2–3)
F. cuvieri	African Hobby	C (2–3)
F. severus	Oriental Hobby	C (3–4)
F. longipennis	Little Falcon	C (2–3)

F. eleonorae	Eleonora's Falcon	A [1–4 (2–3)]
F. concolor	Sooty Falcon	C [1–4 (2–3)]
F. rufigularis	Bat Falcon	D (2–3)
F. femoralis	Aplomado Falcon	C (2–4)
F. hypoleucos	Grey Falcon	C (2–4)
F. subniger	Black Falcon	C (3–4)
F. biarmicus	Lanner Falcon	B (3–4)
F. mexicanus	Prairie Falcon	B [3–6 (4–5)]
F. jugger	Laggar Falcon	C [2–5 (3–4)]
F. kreyenborgi	Kleinschmidt's Falcon	E
F. cherrug	Saker Falcon	B [3–6 (4)]
F. rusticolus	Gyrfalcon	B [2–7 (4)]
F. deiroleucos	Orange-breasted Falcon	D (3?)
F. fasciinucha	Taita Falcon	C (3)
F. peregrinus	Peregrine Falcon	A [2–5 (3–4)]

APPENDIX II
Raptor species occurring in less favourable habitats

All species mentioned breed in the habitat concerned regularly. Status is defined (in this and Appendix III) by the following code letters: R mainly or wholly resident; R/M some resident, some migrant; M mainly or wholly migrant; N nomadic; M/N migrant/nomadic. E and W refer to east and west.

Species	Status	Distribution
TUNDRA		
1 Rough-legged Buzzard *Buteo lagopus*	M	Holarctic
2 Golden Eagle *Aquila chrysaetos*	M	Holarctic
3 Gyrfalcon *Falco rusticolus*	R, R/M	Holarctic
4 Peregrine Falcon *F. peregrinus*	M	Cosmopolitan

Swainson's Hawk, *Buteo swainsonii*, extends to Alaska but is not a typical tundra species; and three northern sea eagles, the White-tailed, *Haliaeetus albicilla*, Bald Eagle, *H. leucocephalus*, and Steller's, *H. pelagicus*, occur in Arctic seas but are not typical tundra birds.

Species	Status	Distribution
TAIGA (Boreal and cold-temperate conifer forest)		
1 Hen Harrier *Circus cyaneus* ⎫ southern	M	Holarctic
2 Pied Harrier *C. melanoleucus* ⎭ fringes only	M	Palearctic
3 Goshawk *Accipiter gentilis*	R, R/M	Holarctic
4 Sparrowhawk *A. nisus*	M, R/M	Palearctic
5 Sharp-shinned Hawk *A. striatus*	M	Nearctic
6 Common Buzzard *Buteo buteo (vulpinus* and *japonicus)*	M	Palearctic
7 Red-tailed Hawk *B. jamaicensis*	M	Nearctic
8 Rough-legged Buzzard *B. lagopus*	M	Holarctic
9 Golden Eagle *Aquila chrysaetos*	R, R/M	Holarctic
10 Merlin *Falco columbarius*	M	Holarctic
11 Gyrfalcon *F. rusticolus*	R	Holarctic
12 Peregrine Falcon *F. peregrinus*	M	Cosmopolitan

Species 1, 2, 4, 5, and 7 breed mainly in southern parts of this zone.

Species	Status	Distribution
TEMPERATE DECIDUOUS WOODLANDS		
1 Turkey Vulture *Cathartes aura*	M	Nearctic
2 Honey Buzzard *Pernis apivorus*	M	Palearctic
3 Black Kite *Milvus migrans*	M, N	Palearctic (Australia)
4 Red Kite *M. milvus*	M	Palearctic
5 Goshawk *Accipiter gentilis*	R	Holarctic
6 Japanese Lesser Sparrowhawk *A. gularis*	M	Palearctic (E)
7 Collared Sparrowhawk *A. cirrhocephalus*	R	Australia
8 Sparrowhawk *A. nisus*	R, R/M	Palearctic
9 Sharp-shinned Hawk *A. striatus*	R, R/M	Nearctic
10 Australian Goshawk *A. fasciatus*	R	Australia
11 Variable Goshawk *A. novaehollandiae*	R	Australia

12	Cooper's Hawk *A. cooperi*	R/M	Nearctic
13	Grey-faced Buzzard Eagle *Butastur indicus*	M	Palearctic (E)
14	Broad-winged Hawk *Buteo platypterus*	M	Nearctic
15	Red-shouldered Hawk *B. lineatus*	R/M	Nearctic
16	Red-tailed Buzzard *B. ventralis*	R	South America
17	Red-tailed Hawk *B. jamaicensis*	R	Nearctic
18	Common Buzzard *B. buteo*	R	Palearctic
19	Lesser Spotted Eagle *Aquila pomarina*	M	Palearctic
20	Greater Spotted Eagle *A. clanga*	M	Palearctic
*21	Mountain Hawk-eagle *Spizaetus nipalensis*	R	Palearctic (Japan)
22	Darwin's Caracara *Phalcoboenus albogularis*	R	South America
23	Hobby *Falco subbuteo*	M	Palearctic

* not best represented in this zone.

TEMPERATE MOORLANDS, MOUNTAINS, AND MARSHES

1	Andean Condor *Vultur gryphus*	R	South America
2	Marsh Harrier *Circus aeruginosus*	M	Palearctic
3	Hen Harrier *C. cyaneus*	R/M	Holarctic
4	Cinereous Harrier *C. cinereus*	R	South America
5	Montagu's Harrier *C. pygargus*	M	Palearctic (W)
6	Pied Harrier *C. melanoleucus*	M	Palearctic (E)
7	Red-backed Buzzard *Buteo polyosoma*	R	South America
8	Common Buzzard *B. buteo*	R	Palearctic
9	Golden Eagle *Aquila chrysaetos*	R	Holarctic
10	Wedge-tailed Eagle *A. audax*	R	Australia
11	Darwin's Caracara *Phalcoboenus albogularis*	R	South America
12	Forster's Caracara *P. australis*	R	South America
13	Common Caracara *Polyborus plancus*	R	South America
14	Chimango *Milvago chimango*	R	South America
15	American Kestrel (Sparrowhawk) *Falco sparverius*	R, R/M	Nearctic
16	Common Kestrel *F. tinnunculus*	R, R/M	Palearctic
17	Nankeen Kestrel *F. cenchroides*	R	Australia
18	Merlin (Pigeon Hawk) *F. columbarius*	M	Holarctic
19	Kleinschmidt's Falcon *F. kreyenborgi*	R?	South America
20	Peregrine Falcon *F. peregrinus*	R	Cosmopolitan

TROPICAL MONTANE MOORLANDS AND MOUNTAINS

1	Andean Condor *Vultur gryphus*	R	South America
2	Gurney's Buzzard *Buteo poecilochrous*	R	South America
3	Curunculated Caracara *Phalcoboenus carunculatus*	R	South America
4	Mountain Caracara *Phalcoboenus megalopterus*	R?	South America

Other species regularly occurring in such habitat include

1	Red-backed Buzzard *Buteo polyosoma*	R	South America
2	Augur Buzzard *B. augur*	R	Africa
3	Grey Eagle-buzzard *Geranoaetus melanoleucus*	R	South America
4	Lammergeier *Gypaetus barbatus*	R	Africa
5	Verreaux's Eagle *Aquila verreauxi*	R	Africa
6	Common Caracara *Polyborus plancus*	R	South America
7	Lanner Falcon *Falco biarmicus*	R	Africa
8	Peregrine Falcon *F. peregrinus*	R	Africa, and elsewhere

and possibly others

Species		Status	Other main habitats	Distribution
DESERTS				
1	Letter-winged Kite *Elanus scriptus*	R/N	Subtropical/Tropical savannas	Australia
2	Egyptian Vulture *Neophron percnopterus*	R	Subtropical/Tropical savannas	Eurasia/Africa
3	European Griffon *Gyps fulvus*	R	Subtropical mountains	Europe
4	Rüppell's Griffon *G. rueppellii*	R	Tropical savannas	Africa
5	Cape Vulture *G. coprotheres*	R	Subtropical mountains	Africa
6	Black Vulture *Aegypius monachus*	R	Subtropical woodlands	Eurasia
7	Lappet-faced Vulture *A. tracheliotus*	R	Tropical savannas	Africa
8	White-headed Vulture *A. occipitalis*	R	Tropical savannas	Africa

9	Serpent Eagle *Circaetus gallicus* (races)	R/M	Subtropical woodlands	Eurasia
0	Dark Chanting Goshawk *Melierax metabates*	R	Tropical savannas	Africa/Arabia
1	Pale Chanting Goshawk *M. canorus*	R	Tropical savannas	Africa
2	Long-legged Buzzard *Buteo rufinus*	R/N	Subtropical steppe	Eurasia
3	Upland Buzzard *B. hemilasius*	R/N	Subtropical steppe	Asia
4	Ferruginous Hawk *B. regalis*	R/N	Subtropical steppe	North America
5	Red-tailed Hawk *B. jamaicensis*	R	Temperate woodlands	North America
6	Steppe/Tawny Eagle *Aquila rapax* (races)	R/M	Subtropical steppe/Tropical savannas	Eurasia/Africa
7	Golden Eagle *A. chrysaetos*	R	Temperate mountains	Eurasia/Africa
8	Wedge-tailed Eagle *A. audax*	R/N	Subtropical steppe/Tropical savannas	Australia
9	Verreaux's Eagle *A. verreauxi*	R	Tropical savannas	Africa
0	Bonelli's Eagle *Hieraaetus fasciatus*	R	Subtropical mountains	Eurasia
1	Martial Eagle *Polemaetus bellicosus*	R	Tropical savannas	Africa
2	Common Caracara *Polyborus plancus*	R/N	Steppe/Tropical savannas	South and Central America
3	Pygmy Falcon *Poliohierax semitorquatus*	R	Tropical savannas	Africa
4	Common Kestrel *Falco tinnunculus*	R/N	Temperate steppes, and elsewhere	Eurasia
5	American Kestrel *F. sparverius*	R/N	Temperate steppes, and elsewhere	America
6	Nankeen Kestrel *F. cenchroides*	R/N	Subtropical steppes, and elsewhere	Australia
7	Brown Hawk *F. berigora*	R/N	Subtropical woodlands, and elsewhere	Australia
8	Red-necked Falcon *F. chicquera*	R	Tropical savannas	Africa
9	Sooty Falcon *F. concolor*	M	Aquatic (Red Sea islands)	Africa
0	Lanner Falcon *F. biarmicus*	R	Tropical savannas	Africa
1	Prairie Falcon *F. mexicanus*	R	Subtropical steppes	North America
2	Saker Falcon *F. cherrug*	R/M	Subtropical steppes	Asia
3	Peregrine Falcon *F. peregrinus*	R/M	Subtropical steppes/Tropical savannas	Cosmopolitan

	Main original habitats		Distribution	
Some, such as the European Snake Eagle, have resident and migrant populations.	Subtropical steppes	10	Cosmopolitan	1
	Subtropical woodlands and mountains	8	Eurasia	10
Some, such as the Honey Buzzard, harriers, Red-footed Falcon, and Lesser Kestrel, migrate across but do not breed in deserts.	Tropical savannas	16	Africa	12
	Temperate areas	4	America	5
	Aquatic	1	Australia	4
	(Some more than one main habitat.)		(Some more than one main region.)	

Species	Distribution

SPECIALIZED HABITATS
Aquatic

Osprey *Pandion haliaetus*	Cosmopolitan
Snail Kite *Rostrhamus sociabilis*	North and South America
Slender-billed Kite *R. hamatus*	South America
Brahminy Kite *Haliastur indus*	Oriental/Australia
White-bellied Sea Eagle *Haliaeetus leucogaster*	Oriental/Australia
Sanford's Sea Eagle *H. sanfordi*	Solomon Islands
African Fish Eagle *H. vocifer*	Africa
Madagascar Fish Eagle *H. vociferoides*	Madagascar
Pallas' Sea Eagle *H. leucoryphus*	Eurasia
Bald Eagle *H. leucogaster*	North America
European Sea Eagle *H. albicilla*	Eurasia
Steller's Sea Eagle *H. pelagicus*	North-east Asia
Lesser Fishing Eagle *Ichthyophaga nana*	Oriental
Grey-headed Fishing Eagle *I. ichthyaetus*	Oriental
Vulturine Fish Eagle *Gypohierax angolensis*	Africa
Marsh Harrier *Circus aeruginosus*	All Old World
African Marsh Harrier *C. ranivorus*	Africa
Long-winged Harrier *C. buffoni*	South America
Common Black Hawk *Buteogallus anthracinus*	North and South America
Rufous Crab Hawk *B. aequinoctialis*	South America
Great Black Hawk *B. urubitinga*	South America
Savanna Hawk *Heterospizias meridionalis*	South America

23	Fishing Buzzard *Busarellus nigricollis*					South America
24	Eleonora's Falcon *Falco eleonorae*					Mediterranean
*25	Sooty Falcon *F. concolor*					Red Sea

In addition, several other species such as the Grey Frog Hawk, *Accipiter soloensis*, Red-shouldered Hawk, *Buteo lineatus*, and Whistling Eagle, *Haliastur sphenurus*, tend to take much prey in marshy areas; the Black Kite, *Milvus migrans*, commonly scavenges in harbours and the Red Kite, *Milvus milvus*, does so in the Cape Verde Islands.

* not confined to this habitat.

Species	Distribution
Towns (Regular, common scavengers breeding in towns and villages)	
1 American Black Vulture *Coragyps atratus*	Central and South America
2 Black Kite *Milvus migrans*	All Old World
3 Egyptian Vulture *Neophron percnopterus*	Asia, Africa
4 Hooded Vulture *Necrosyrtes monachus*	Africa
5 Indian White-backed Vulture *Gyps bengalensis*	Asia
6 African White-backed Vulture *G. africanus*	Africa
7 Long-billed Vulture *G. indicus*	Asia
8 Tawny Eagle *Aquila rapax*	Asia, Africa
9 Common Caracara *Polyborus plancus*	South America
10 Chimango *Milvago chimango*	South America
Towns (Species attracted to towns because of special food or breeding sites)	
1 Bat Hawk *Machaerhamphus alcinus* (bats)	Asia, Africa
2 Yellow-headed Caracara *Milvago chimachima* (scavenger)	South America
3 Lesser Kestrel *Falco naumanni* (breed in buildings)	Europe, North Africa
4 American Kestrel *F. sparverius* (breed in buildings)	Americas
5 Common Kestrel *F. tinnunculus* (breed in buildings)	Eurasia, Africa
6 Madagascar Kestrel *F. newtoni* (breed in buildings)	Madagascar
7 Seychelles Kestrel *F. araea* (breed in buildings)	Seychelles
8 Molucca Kestrel *F. moluccensis* (breed in buildings)	Moluccas
9 Bat Falcon *F. rufigularis* (breed in buildings)	South America
10 Lanner Falcon *F. biarmicus* (breed in buildings)	Africa
11 Laggar Falcon *F. jugger* (breed in buildings)	Asia, India
12 Orange-breasted Falcon *F. deiroleucos* (breed in buildings)	South America and Central America
13 Peregrine Falcon *F. peregrinus* (breed in buildings)	Cosmopolitan

Note 1 Of ten regular species, the Tawny Eagle breeds in towns in Asia and North Africa but apparently not in South Africa or in the steppes of Europe.
2 Of thirteen species attracted to towns, eleven are attracted to breed in buildings, all falcons seeking artificial cliffs. Of these, the Lesser Kestrel is the most attracted, frequently breeding in cathedral towers in Spain and other places.

APPENDIX III
Species occurring in four richest habitats

Note 1 subtropical woodlands and mountains; **2** temperate/subtropical open steppes and grasslands; **3** tropical savannas; **4** tropical forests.
* genus confined to tropical forest. † preferred habitat. O occurs but not main habitat. R resident. RM resident/migrant. M migrant. N nomadic.

Species	1	2	3	4	Main distribution
Turkey Vulture *Cathartes aura*	†RM	†RM	†R		Americas
Yellow-headed Vulture *C. burrovianus*			†R		South America
Greater Yellow-headed Vulture *C. melambrotus*			†R	OR	South America
Black Vulture *Coragyps atratus*	OR	OR	†R		Americas
*King Vulture *Sarcorhamphus papa*				†R	South and Central America
California Condor *Gymnogyps californianus*	†R				North America (W)
Andean Condor *Vultur gryphus*	†R				South America
African Cuckoo Falcon *Aviceda cuculoides*			OR	†R	Africa

Species					Region
Madagascar Cuckoo Falcon *A. madagascariensis*			OR	†R	Madagascar
Jerdon's Baza *A. jerdoni*				†R	Oriental
Crested Baza *A. subcristata*			OR	†R	Australasia
Black Baza *A. leuphotes*		†M	OM/N		Oriental
★Cayenne Kite *Leptodon cayanensis*				†R	South America
★Hook-billed Kite *Chondrohierax uncinatus*				†R	Central and South America
★Long-tailed Honey Buzzard *Henicopernis longicauda*				†R	New Guinea and elsewhere
Black Honey Buzzard *H. infuscata*				†R	New Britain
Honey Buzzard *Pernis apivorus*	OM			†R	Old World, Asia, Europe
Barred Honey Buzzard *P. celebensis*				†R	Celebes
American Swallow-tailed Kite *Elanoides forficatus*	†M		†OM	†R	Americas
Bat Hawk *Machaerhamphus alcinus*			R	†R	Asia, Africa
Pearl Kite *Gampsonyx swainsonii*				†R	Central and South America
White-tailed Kite *Elanus leucurus*		†N		†N	North and South America
Black-shouldered Kite *E. caeruleus*		†N		†N	Europe, Asia, Africa
Australian Black-shouldered Kite *E. notatus*		†N		†N	Australia
Letter-winged Kite *E. scriptus*		†N		†N	Australia
African Swallow-tailed Kite *Chelictinia riocourii*			†M		Africa
★Double-toothed Kite *Harpagus bidentatus*				†R	South America
Rufous-thighed Kite *H. diodon*				†R	South America
Plumbeous Kite *Ictinia plumbea*			†R/M	†R/M	South America
Mississippi Kite *I. misisippiensis*		†M			North America
Square-tailed Kite *Lophoictinia isura*		†R		†R	Australia
Black-breasted Buzzard Kite *Hamirostra melanosternon*		†R		†R	Australia
Black Kite *Milvus migrans*	†M	OM	†M	OM	Old World (all)
Red Kite *M. milvus*	†R/M				Europe
Whistling Eagle *Haliastur sphenurus*	OR	OR		†R	Australia
Brahminy Kite *H. indus*	OR		†R	OR	Oriental/Australasia
Vulturine Fish Eagle *Gypohierax angolensis*			†R	†R	Africa
Egyptian Vulture *Neophron percnopterus*	†M	†M		†R	Eurasia/Africa
Lammergeier *Gypaetus barbatus*	†R				Eurasia/Africa
Hooded Vulture *Necrosyrtes monachus*			†R	OR	Africa
Indian White-backed Vulture *Gyps bengalensis*				†R	Oriental
African White-backed Vulture *G. africanus*				†R	Africa
Long-billed Vulture *G. indicus*	†R	†R		†R	Oriental
Rüppell's Griffon *G. rueppellii*				†R	Africa
Himalayan Griffon *G. himalayensis*	†R				Oriental
Griffon Vulture *G. fulvus*	†R/M				Europe
Cape Vulture *G. coprotheres*	†R		OR		Africa
Lappet-faced Vulture *Aegypius tracheliotus*		†R	†R		Africa
Indian King Vulture *A. calvus*			†R	OR	Oriental
Black Vulture *A. monachus*	†R/M				Eurasia
White-headed Vulture *A. occipitalis*			†R		Africa
Serpent Eagle, European Snake Eagle *Circaetus gallicus* (including Beaudouin's and Black-breasted as races)	†M	OM		†R	Eurasia/Africa
Brown Snake Eagle *C. cinereus*				†R	Africa
Southern Banded Snake Eagle *C. fasciolatus*			†R	†R	Africa
Smaller Banded Snake Eagle *C. cinerascens*				†R	Africa
Bateleur *Terathopius ecaudatus*			†R/N		Africa
Philippine Serpent Eagle *Spilornis holospilus*			†R	†R	Philippines
Celebes Serpent Eagle *S. rufipectus*				†R	Celebes
Crested Serpent Eagle *S. cheela*			OR	†R	Oriental
Nicobar Serpent Eagle *S. klossi*				†R	Nicobar Islands
Andaman Serpent Eagle *S. elgini*				†R	Andaman Islands
★Congo Serpent Eagle *Dryotriorchis spectabilis*				†R	Africa
★Madagascar Serpent Eagle *Eutriorchis astur*				†R	Madagascar
African Harrier Hawk *Polyboroides typus*			OR	†R	Africa
Madagascar Harrier Hawk *P. radiatus*			†R	OR	Madagascar
Crane Hawk *Geranospiza caerulescens*			†R	OR	South America
Spotted Harrier *Circus assimilis*	†R/M		†R		Australia
Marsh Harrier *C. aeruginosus*	†R/M		†R		Cosmopolitan excluding America
African Marsh Harrier *C. ranivorus*	†R		†R		Africa
Black Harrier *C. maurus*	†R				South Africa
Cinereous Harrier *C. cinereus*	†R/M		†R		South America
Pale Harrier *C. macrourus*	†M				Eurasia
Montagu's Harrier *C. pygargus*	†M				Europe
Long-winged Harrier *C. buffoni*	†R		†R		South America

Species	1	2	3	4	Distribution
Dark Chanting Goshawk *Melierax metabates*		OR	†R		Africa, Arabia
Pale Chanting Goshawk *M. canorus*		†R	†R		Africa
Gabar Goshawk *M. gabar*			†R		Africa
*Doria's Goshawk *Megatriorchis doriae*				†R	Australasia
Red Goshawk *Erythrotriorchis radiatus*	OR		†R		Australasia
Goshawk *Accipiter gentilis*	OR				Eurasia/North America
Henst's Goshawk *A. henstii*			OR	†R	Madagascar
Black Sparrowhawk *A. melanoleucus*			OR	†R	Africa
Meyer's Goshawk *A. meyerianus*				†R	Oriental/Australasia
Buerger's Goshawk *A. buergersi*				†R	New Guinea
Ovampo Sparrowhawk *A. ovampensis*			†R/M		Africa
Madagascar Sparrowhawk *A. madagascariensis*			†R		Madagascar
Japanese Lesser Sparrowhawk *A. gularis*	†M				East Asia
Besra Sparrowhawk *A. virgatus*			OR	†R	Oriental
Celebes Little Sparrowhawk *A. nanus*				†R	Celebes
Vinous-breasted Sparrowhawk *A. rhodogaster*				†R	Celebes
Moluccan Sparrowhawk *A. erythrauchen*				†R	Moluccas
Collared Sparrowhawk *A. cirrhocephalus*	†R		†R		Australasia
New Britain Sparrowhawk *A. brachyurus*				†R	New Britain
European Sparrowhawk *A. nisus*	†R				Eurasia
Rufous-breasted Sparrowhawk *A. rufiventris*	†R			†R	Africa
Sharp-shinned Hawk *A. striatus*	†R				Americas
Red-thighed Sparrowhawk *A. erythropus*				†R	Africa
African Little Sparrowhawk *A. minullus*			†R	†R	Africa
Chestnut-bellied Sparrowhawk *A. castanilius*				†R	Africa
African Goshawk *A. tachiro*			OR	†R	Africa
Crested Goshawk *A. trivirgatus*				†R	Oriental
Celebes Crested Goshawk *A. griseiceps*				†R	Celebes
Spot-tailed Accipiter *A. trinotatus*				†R	Celebes
Blue and Grey Sparrowhawk *A. luteoschistaceus*				†R	New Britain
Australian Goshawk *A. fasciatus*	†R		†R		Australia
Gray's Goshawk *A. henicogrammus*			†R		Moluccas
Variable Goshawk *A. novaehollandiae*	†R		†R	†R	Australasia
Grey-throated Goshawk *A. griseogularis*				†R	Moluccas
Black-mantled Accipiter *A. melanochlamys*				†R	New Guinea
Imitator Sparrowhawk *A. imitator*				†R	Solomon Islands
Pied Goshawk *A. albogularis*				†R	Solomon Islands
New Caledonia Sparrowhawk *A. haplochrous*				†R	New Caledonia
Fiji Goshawk *A. rufitorques*			OR	†R	Fiji Islands
New Guinea Grey-headed Goshawk *A. poliocephalus*				†R	New Guinea
New Britain Grey-headed Goshawk *A. princeps*				†R	New Britain
Grey Frog Hawk *A. soloensis*	†M				East Asia
Levant Sparrowhawk *A. brevipes*	†M				Eurasia
Shikra *A. badius*			†R/M		India/Africa
Nicobar Shikra *A. butleri*				†R	Nicobar Islands
France's Sparrowhawk *A. francesii*			†R		Madagascar and elsewhere
American Collared Sparrowhawk *A. collaris*	†R		†R		Americas
Tiny Sparrowhawk *A. superciliosus*				†R	South America
Gundlach's Hawk *A. gundlachi*				†R	Cuba
Cooper's Hawk *A. cooperii*	†R/M				North America
Bicoloured Sparrowhawk *A. bicolor*				†R	South America
Grey-bellied Goshawk *A. poliogaster*				†R	South America
*African Long-tailed Hawk *Urotriorchis macrourus*				†R	Africa
Grasshopper Buzzard-eagle *Butastur rufipennis*			†M		Africa
Rufous-winged Buzzard-eagle *B. liventer*			†R		Oriental
White-eyed Buzzard *B. teesa*			†R		India and elsewhere
Grey-faced Buzzard-eagle *B. indicus*	†M				East Asia
Lizard Buzzard *Kaupifalco monogrammicus*			†R		Africa
Slate-coloured Hawk *Leucopternis schistacea*				†R	South America
Plumbeous Hawk *L. plumbea*				†R	South America
Prince's Hawk *L. princeps*				†R	South America
Black-faced Hawk *L. melanops*				†R	South America
White-browed Hawk *L. kuhli*				†R	South America
White-necked Hawk *L. lacernulata*				†R	South America
Semiplumbeous Hawk *L. semiplumbea*				†R	South America
White Hawk *L. albicollis*			OR	†R	South and Central America

Grey-backed Leucopternis *L. occidentalis*			†R		North and South America
Mantled Hawk *L. polionota*			†R		South America
Black Solitary Eagle *Harpyhaliaetus solitarius*			†R		South America
Crowned Solitary Eagle *H. coronatus*		†R			South America
Savanna Hawk *Heterospizias meridionalis*		†R	†R		South America
Grey Eagle-buzzard *Geranoaetus melanoleucus*		†R			South America
Bay-winged Hawk *Parabuteo unicinctus*	†R	OR	†R		Central and South America
Grey Hawk *Buteo nitidus*	†R		†R		Central and South America
Roadside Hawk *B. magnirostris*	†R		†R		Central and South America
Rufous-thighed Hawk *B. leucorrhous*			†R		South America
Ridgway's Hawk *B. ridgwayi*			†R		South America
Broad-winged Hawk *B. platypterus*			OR		Central America and West Indies
Short-tailed Hawk *B. brachyurus*	†R	†R	†R		Central and South America
Red-shouldered Hawk *B. lineatus*	†R				Central America
Zone-tailed Hawk *B. albonotatus*	†R		†R		Central and South America
Hawaiian Hawk *B. solitarius*	†R				Hawaii
Swainson's Hawk *B. swainsonii*		†M			North America
Galápagos Hawk *B. galapagoensis*			†R		Galápagos Islands
White-tailed Hawk *B. albicaudatus*	†R	OR	†R		Central and South America
Red-backed Buzzard *B. polyosoma*		†R			South America
Red-tailed Hawk *B. jamaicensis*	†R	OR			North America
Common Buzzard *B. buteo*	†R				Eurasia
African Mountain Buzzard *B. oreophilus*	†R		†R		Africa
Madagascar Buzzard *B. brachypterus*			†R		Madagascar
Long-legged Buzzard *B. rufinus*	†R	†R			Eurasia/Africa
Upland Buzzard *B. hemilasius*		†R			Asia
Ferruginous Hawk *B. regalis*		†R			North America
African Red-tailed Buzzard *B. auguralis*			†M		Africa
Jackal Buzzard *B. rufofuscus*	†R	†R			South Africa
Augur Buzzard *B. augur*			†R		Africa
*Guiana Crested Eagle _Morphnus guianensis_			†R		South America
*Harpy Eagle _Harpia harpyja_			†R		South America
*New Guinea Harpy Eagle _Harpyopsis novaeguineae_			†R		New Guinea
*Philippine Monkey-eating Eagle _Pithecophaga jefferyi_			†R		Philippines
*Indian Black Eagle _Ictinaetus malayensis_			†R		Oriental
Tawny/Steppe Eagle *Aquila rapax*		†M	†R		Asia/Africa
Imperial Eagle *A. heliaca*	†R	OM			Europe
Wahlberg's Eagle *A. wahlbergi*			†M		Africa
Gurney's Eagle *A. gurneyi*			†R		New Guinea
Golden Eagle *A. chrysaetos*	†R				Europe/America
Wedge-tailed Eagle *A. audax*	†R	†R/N	†R/N		Australia
Verreaux's Eagle *A. verreauxi*	†R		†R		Africa
Bonelli's Eagle, African Hawk-eagle *Hieraaetus fasciatus*	†R		†R		Europe/Africa
Booted Eagle *H. pennatus*	†M				Europe/Africa
Little Eagle *H. morphnoides*	†R		†R	OR	Australasia
Ayres' Hawk-eagle *H. dubius*			OR	†R	Africa
Chestnut-bellied Hawk-eagle *H. kienerii*			†R		Oriental
*Black and White Hawk-eagle _Spizastur melanoleucus_			†R		Central and South America
Long-crested Eagle *Spizaetus occipitalis*		†R	OR	†R	Africa
Cassin's Hawk-eagle *S. africanus*			†R		Africa
Changeable Hawk-eagle *S. cirrhatus*		†R	†R		Oriental
Mountain Hawk-eagle *S. nipalensis*	†R		†R		Oriental
Java Hawk-eagle *S. bartelsi*			†R		Java
Celebes Hawk-eagle *S. lanceolatus*			†R		Celebes
Philippine Hawk-eagle *S. philippensis*			†R		Philippines
Blyth's Hawk-eagle *S. alboniger*			†R		Oriental
Wallace's Hawk-eagle *S. nanus*			†R		Oriental
Black Hawk-eagle *S. tyrannus*			†R	OR	Central and South America
Ornate Hawk-eagle *S. ornatus*			†R		Central and South America
Crowned Eagle *Stephanoaetus coronatus*			OR	†R	Africa
*Isidore's Eagle _Oroaetus isidori_			†R		South America
Martial Eagle *Polemaetus bellicosus*	OR	OR	†R		Africa
Secretary Bird *Sagittarius serpentarius*		†R	†R		Africa
Yellow-throated Caracara *Daptrius ater*			†R		South America
Red-throated Caracara *D. americanus*			OR	†R	South America
Mountain Caracara *Phalcoboenus megalopterus*	†R				South America
Common Caracara *Polyborus plancus*	†R	†R	†R		Central and South America

Species	1	2	3	4	Distribution
Chimango *Milvago chimango*		†R			South America
Yellow-headed Caracara *M. chimachima*			†R		South America
*Laughing Falcon *Herpetotheres cachinnans*				†R	Central and South America
*Barred Forest Falcon *Micrastur ruficollis*				†R	Central and South America
Sclater's Forest Falcon *M. plumbeus*				†R	South America
Slaty-backed Forest Falcon *M. mirandollei*				†R	Central and South America
Collared Forest Falcon *M. semitorquatus*				†R	Central and South America
Traylor's Forest Falcon *M. buckleyi*				†R	Central and South America
Spot-winged Falconet *Spiziapteryx circumcinctus*			†R		South America
Pygmy Falcon *Poliohierax semitorquatus*		†R	†R		Africa
Fielden's Falconet *P. insignis*			†R		Oriental
*Red-legged Falconet *Microhierax caerulescens*				†R	Oriental
Black-legged Falconet *M. fringillarius*				†R	Oriental
Bornean Falconet *M. latifrons*				†R	Borneo
Philippine Falconet *M. erythrogonys*				†R	Philippines
Pied Falconet *M. melanoleucus*				†R	Oriental
Lesser Kestrel *Falco naumanni*	†M	†M			Europe/Asia
Greater Kestrel *F. rupicoloides*		†R	†R		Africa
Fox Kestrel *F. alopex*			†R		Africa
American Kestrel *F. sparverius*	†R	†R	†R		Americas
Common Kestrel *F. tinnunculus*	†R	†R/N	†R		Eurasia/Africa
Madagascar Kestrel *F. newtoni*			†R		Madagascar
Mauritius Kestrel *F. punctatus*				†R	Mauritius
Seychelles Kestrel *F. araea*			†R	†R	Seychelles
Molucca Kestrel *F. moluccensis*			†R		Moluccas
Nankeen Kestrel *F. cenchroides*	†R	†R	†R		Australia
Grey Kestrel *F. ardosiaceus*			†R		Africa
Dickinson's Kestrel *F. dickinsoni*			†R		Africa
Madagascar Banded Kestrel *F. zoniventris*			†R		Madagascar
Red-footed Falcon *F. vespertinus*	†M				Europe/Asia
Red-headed Falcon *F. chicquera*		OR	†R		Africa/India
Merlin *F. columbarius*		†R			Eurasia/North America
Brown Hawk *F. berigora*	†R	†R/N	†R/N		Australia
New Zealand Falcon *F. novaezeelandiae*	†R				New Zealand
Hobby *F. subbuteo*	†M				Eurasia
African Hobby *F. cuvieri*	†R		†R	†R	Africa
Oriental Hobby *F. severus*				†R	Oriental
Little Falcon *F. longipennis*	†R			†R	Australasia
Bat Falcon *F. rufigularis*			OR	†R	South America
Aplomado Falcon *F. femoralis*		†R/M	†R		Americas
Grey Falcon *F. hypoleucos*	OR	†R/N	†R		Australia
Black Falcon *F. subniger*	†R	OR	†R		Australia
Lanner Falcon *F. biarmicus*	†R	†R	†R		Eurasia/Africa
Laggar Falcon *F. jugger*	†R		†R		Oriental
Prairie Falcon *F. mexicanus*	†R	†R			North America
Saker Falcon *F. cherrug*	†R/M	†M			Eurasia
Orange-breasted Falcon *F. deiroleucos*			OR	†R	South America
Taita Falcon *F. fasciinucha*			†R		Africa
Peregrine Falcon *F. peregrinus*	†R	†R/M	†R	OR	Cosmopolitan

Richest habitats

1 Tropical savannas

112	Preferred habitat
18	Occur
130	Species

2 Tropical forest

111	Preferred habitat
13	Occur
124	Species

3 Subtropical wooded areas and mountains

65	Preferred habitat
8	Occur
73	Species

4 Subtropical/Temperate plains/Steppe

47 Preferred habitat
12 Occur
59 Species

Distribution	1	2	3	4
Cosmopolitan	1	1	1	1
Eurasia (Palearctic)	25	16		
Africa (Ethiopian)	15	19	61	28
North America (Nearctic)	15	12		
Central and South America (Neotropical)	16	14	34	40
Tropical Asia (Oriental)	9	3	25	38
Australia (Australasian)	15	12	20	20

APPENDIX IV
Migrant and nomadic species

Note 1 Main breeding range or wintering ranges are Arctic, temperate (TE), subtropical (STR), tropical (TR), or combinations.
2 Less than 300 grams = small; 300 to 600 grams = medium/small; 600 to 1000 grams = medium/large; 1000 to 2000 grams = large; over 2000 grams = very large.
3 N and S refer to northern and southern hemispheres.

Species	Range (Note 1)		Weight grams (Note 2)
	Breeding	Winter	
A Complete migrants			
1 African Swallow-tailed Kite *Chelictinia riocourii*	TR	TR	small
2 Mississippi Kite *Ictinia misisippiensis*	TE/STR	TR	small
3 Black Kite *Milvus migrans*	TE/STR	TR	medium/large
4 Pale Harrier *Circus macrourus*	TE/STR	TR	medium/small
5 Montagu's Harrier *C. pygargus*	TE/STR	TR	medium/small
6 Pied Harrier *C. melanoleucus*	TE/STR	TR	medium/small
7 Japanese Lesser Sparrowhawk *Accipiter gularis*	TE	TR	small
8 Grey Frog Hawk *A. soloensis*	TE	TR	small
9 Levant Sparrowhawk *A. brevipes*	STR	STR/TR	small
10 Grasshopper Buzzard-eagle *Butastur rufipennis*	TR	TR	medium/small
11 Grey-faced Buzzard-eagle *B. indicus*	TE/STR	TR	medium/small
12 Swainson's Hawk *Buteo swainsonii*	TE/STR(N)	TE/STR(S)	medium/large
13 Rough-legged Buzzard *B. lagopus*	Arctic	TE	medium/large
14 African Red-tailed Buzzard *B. auguralis*	TR	TR	medium/large
15 Lesser Spotted Eagle *Aquila pomarina*	TE/STR	TR(S)	large
16 Greater Spotted Eagle *A. clanga*	TE/STR	TR(N)	large
17 Wahlberg's Eagle *A. wahlbergi*	TR(S)	TR(N?)	medium/large
18 Booted Eagle *Hieraaetus pennatus*	STR	TR	medium/large
19 Lesser Kestrel *Falco naumanni*	STR	TR/STR(S)	small
20 Red-footed Falcon *F. vespertinus*	TE/STR	TR/STR(S)	small
21 Hobby *F. subbuteo*	TE/STR	TR(S)	small
22 Eleonora's Falcon *F. eleonorae*	STR	TR	small
23 Sooty Falcon *F. concolor*	TR	TR	small
Arctic	1	0	10 small
TE	2	1	5 medium/small
TE/STR/TR	11	4	6 medium/large
STR	4		2 large
TR	4	18	
B Mainly migrant species			
1 Osprey *Pandion haliaetus*	TE/TR	TE/TR	large
2 Honey Buzzard *Pernis apivorus*	TE/TR	TR	medium/large
3 American Swallow-tailed Kite *Elanoides forficatus*	STR/TR	TR?	medium/small
4 Plumbeous Kite *Ictinia plumbea*	TR	TR	small
5 Red Kite *Milvus milvus*	TE/STR	STR	medium/large
6 Marsh Harrier *Circus aeruginosus*	TE/STR/TR	TR/STR	medium/small
7 Sharp-shinned Hawk *Accipiter striatus*	TE	TE/STR	small
8 Broad-winged Hawk *Buteo platypterus*	TE/STR	STR/TR	medium/small
9 Steppe/Tawny Eagle *Aquila rapax*	TE/STR	TR	very large
10 Merlin *Falco columbarius*	TE/STR	TE/STR	small

11	Saker Falcon *F. cherrug*	TE/STR	STR/TR	medium/large
*12	Peregrine Falcon *F. peregrinus*	Arctic/TE/STR	TE/TR	medium/large

Arctic	1 (partial)	0	3 small
TE	1	0	3 medium/small
TE/STR/TR	9	5	4 medium/large
STR/TR	1	3	1 large
TR	1	4	1 very large

* Peregrine might be regarded as partly rather than mainly migratory but probably the largest populations migrate.

Species	Range		Weight
	Breeding	**Winter**	
C Partly migrant species			
1 Turkey Vulture *Cathartes aura*	TE/STR	STR/TR	large
2 African Cuckoo Falcon *Aviceda cuculoides*	TR	TR	small
3 Black Baza *A. leuphotes*	TR	TR	small
4 Pallas' Sea Eagle *Haliaeetus leucoryphus*	TE/STR	STR/TR	very large
5 Bald Eagle *H. leucocephalus*	Arctic/TE/TR	TE/STR	very large
6 White-tailed Eagle *H. albicilla*	Arctic/TE	TE	very large
7 Steller's Sea Eagle *H. pelagicus*	Arctic/TE	TE	very large
8 Egyptian Vulture *Neophron percnopterus*	STR/TR	TR	large
9 European Griffon *Gyps fulvus*	STR	STR	very large
10 Black Vulture *Aegypius monachus*	STR	STR/TR	very large
11 European Snake Eagle *Circaetus gallicus*	STR	TR	large
12 Spotted Harrier *Circus assimilis*	STR/TR	STR/TR	medium/small
13 Hen Harrier *C. cyaneus*	Arctic/TE	TE	medium/small
14 Cinereous Harrier *C. cinereus*	STR	STR/TR?	medium/small
15 Dark Chanting Goshawk *Melierax metabates*	STR	TR	medium/small
16 Goshawk *Accipiter gentilis*	Arctic/TE/STR	TE/STR	medium/large
17 Ovampo Sparrowhawk *A. ovampensis*	TR	TR	small
18 European Sparrowhawk *A. nisus*	TE/STR	TE/TR	small
19 Shikra *A. badius*	TR	TR	small
20 Cooper's Hawk *A. cooperii*	TE/STR	STR	medium/small
21 Grey Hawk *Buteo nitidus*	STR/TR	TR	medium/small
22 Red-shouldered Hawk *B. lineatus*	TE/STR	STR	medium/large
23 White-tailed Hawk *B. albicaudatus*	STR/TR	TR	medium/large
24 Red-backed Buzzard *B. polyosoma*	TE/STR(S)	STR(S)	large
25 Zone-tailed Hawk *B. albonotatus*	STR/TR	STR/TR	medium/large
26 Red-tailed Hawk *B. jamaicensis*	TE/STR	TE/TR	large
27 Common Buzzard *B. buteo*	TE/STR	TE/TR	medium/large
28 Long-legged Buzzard *B. rufinus*	STR(TE?)	STR/TR	large
29 Upland Buzzard *B. hemilasius*	STR	STR	large
30 Ferruginous Hawk *B. regalis*	TE/STR	STR	large
31 Imperial Eagle *Aquila heliaca*	STR	STR/TR	very large
32 Golden Eagle *A. chrysaetos*	Arctic/STR	TE/STR	very large
33 Bonelli's Eagle *Hieraaetus fasciatus*	STR/TR	STR/TR	large
34 American Kestrel *Falco sparverius*	TE/STR	STR/TR	small
35 Common Kestrel *F. tinnunculus*	TE/TR	TE/TR	small
36 Little Falcon *F. longipennis*	STR/TR	STR/TR	small
37 Aplomado Falcon *F. femoralis*	STR/TR	TR?	small
38 Prairie Falcon *F. mexicanus*	TE/STR	STR	medium/large
39 Gyrfalcon *F. rusticolus*	Arctic	Arctic/TE	large

Arctic	1(+ 5 partly)	1 (partly)	
TE	4 (mainly	3	9 small
TE/STR/TR	13	7	6 medium/small
STR	7	7	6 medium/large
STR/TR	8	11	10 large
TR	4	10	8 very large

D Typically nomadic species (some also migratory)			
1 Black Baza *Aviceda leuphotes*		TR	small
2 White-tailed Kite *Elanus leucurus*		TE/STR/TR	small
3 Black-shouldered Kite *E. caeruleus*		STR/TR	small
4 Australian Black-shouldered Kite *E. notatus*		STR/TR	small

5	Letter-winged Kite *E. scriptus*	STR/TR	small
6	Bateleur *Terathopius ecaudatus*	TR	large
7	Spotted Harrier *Circus assimilis*	STR/TR	medium/small
8	Tawny Eagle *Aquila rapax*	TR (races)	large
9	Wedge-tailed Eagle *A. audax*	TE/TR	very large
10	Secretary Bird *Sagittarius serpentarius*	STR/TR	very large
11	Kestrel *Falco tinnunculus*	TE/TR	small
12	Nankeen Kestrel *F. cenchroides*	TE/TR	small
13	Brown Hawk *F. berigora*	TE/TR	medium/small
14	Little Falcon *F. longipennis*	STR/TR	small
15	Grey Falcon *F. hypoleucus*	STR/TR	medium/large
16	Black Falcon *F. subniger*	STR/TR	medium/large
17	Lanner Falcon *F. biarmicus*	TR	medium/large

TE	None	8 small
TE/STR/TR	5	2 medium/small
STR/TR	8	3 medium/large
TR	4	2 large
		2 very large

Cathartes aura, Coragyps atratus, Vultur gryphus·(locally):
Gyps, Aegypius, Neophron spp (locally); *Phalcoboenus,
Polyborus,* and *Milvago* spp possibly.

APPENDIX V
Threatened species of birds of prey

Species	Numbers	Main threats	Conservation measures
A Acutely threatened			
1 California Condor	40–50	Indirect persecution, human pressures generally	Actively being pursued
2 Madagascar Serpent Eagle	?	Forest destruction	Nothing being done at all
3 Philippine Monkey-eating Eagle	Less than 100	Persecution (? stopped), forest destruction	Active, but not completely effective
4 Mauritius Kestrel	Less than 10	Persecution, forest destruction	Active, including captive breeding
B Less acutely threatened species			
1 Andean Condor	000s?	Persecution	None
2 European Sea Eagle	1–2000? decreasing	Pesticides	Active in areas of decrease (Europe)
3 Bald Eagle	50–60 000 + decreasing	Pesticides: persecution	Active in areas of decrease
4 Steller's Sea Eagle	000s? said to be decreasing	? Little known	Not known
5 Lammergeier	50 000 + ? decreasing in Europe	Improved veterinary sciences and stock keeping	None, except locally where rare, and not effective
6 European Griffon	000s?	Improved stock keeping reducing carrion	None effective except locally
7 Cape Vulture	000s?	Improved stock keeping killed by electric wires	Active study in progress
8 European Black Vulture	00s?	Improved stock keeping reducing carrion	None effective except locally
9 Montagu's Harrier	000s?	Pesticides perhaps in wintering area: persecution	None effective
10 Black Harrier	00s?	Marshland drainage	None effective
11 New Britain Grey-headed Goshawk	0s?	Forest destruction in small range	None
12 Gundlach's Hawk	00s?		
13 Galápagos Hawk	About 130 pairs	Persecution due to tameness	Partially effective protection
14 Lesser Spotted Eagle	000s	All decreasing or threatened by increased human invasion of habitats: pesticides	Active protective measures taken where decreasing: none elsewhere
15 Greater Spotted Eagle	000s		
16 Imperial Eagle	Under 5000?		

17	Seychelles Kestrel	At least 150	Persecution: catastrophe: is extremely tame	Believed stable and persecution decreasing
18	Saker Falcon	ooos	⎫ Demands of falconers: pesticides (but this threat receding): some direct persecution (Peregrine)	Efforts to guard eyries and so on in areas where especially threatened: usually none effective
19	Gyrfalcon	ooos		
20	Peregrine Falcon	ooos		

	Species	Status and knowledge (Is = Islands)	Threats to existence

C Possibly threatened species

	Species	Status and knowledge	Threats to existence
1	Long-tailed Honey Buzzard	Is: unknown: nest unknown	Forest destruction
2	Black Honey Buzzard	Is: oos? unknown: nest unknown	Forest destruction
3	Barred Honey Buzzard	Is: oos? unknown: nest unknown	Forest destruction
4	Sanford's Sea Eagle	Is: oos? unknown: nest unknown	Forest destruction?
5	Philippine Serpent Eagle	Is: ooos? unknown: nest undescribed	? May not be threatened
6	Celebes Serpent Eagle	Is: ooos? unknown: nest little known	Probably not acute
7	Nicobar Serpent Eagle	Is: oos? unknown: breeding unknown	Forest destruction?
8	Andaman Serpent Eagle	Is: oos? unknown: breeding unknown	Habitat destruction?
9	Red Goshawk	Limited habitat: rare: numbers unknown	None obvious
10	Henst's Goshawk	Is: oos? little known	Forest destruction
11	Celebes Little Sparrowhawk	Is: status obscure: nest unknown	Forest destruction
12	Vinous-breasted Sparrowhawk	Is: oos? unknown: nest unknown	Forest destruction
13	New Britain Sparrowhawk	Is: oos? unknown: nest unknown	Forest destruction
14	Celebes Crested Goshawk	Is: little known, more widespread	Forest destruction
15	Spot-tailed Accipiter	Is: little known: virgin forest	Forest destruction
16	Blue and Grey Sparrowhawk	Is: Montane forests: almost unknown	Forest destruction
17	Black-mantled Accipiter	Is: Montane forests	Forest destruction
18	Imitator Sparrowhawk	Is(2): unknown: nest unknown	Forest destruction
19	Pied Goshawk	Is: habits almost unknown	Forest destruction
20	New Caledonia Sparrowhawk	Is: little known	Forest destruction
21	Fiji Goshawk	Is: little known	Forest destruction: may be adaptable
22	Nicobar Shikra	Is: oos? little known	Habitat destruction
23	Plumbeous Hawk	Small range in lowland South America: little known: said to be uncommon	Habitat destruction
24	Grey-backed Hawk	West Ecuador: unknown: nest unknown	Habitat destruction
25	Hawaiian Hawk	Is: wide habitat range	Probably not threatened
26	Gurney's Buzzard	Small Andean range: unknown	Probably not threatened
27	Red-tailed Buzzard	oos? small South American range	Habitat destruction?
28	Mountain Buzzard	Montane east and South Africa	Not immediately threatened
29	Harpy Eagle	Wide ranging, but said to be rare	Forest destruction
30	New Guinea Harpy Eagle	Large range, but probably rare	Forest destruction: some killed for feathers
31	Gurney's Eagle	Large range, but apparently rare and unknown	Forest destruction
32	Java Hawk-eagle	Limited range: almost unknown	Forest destruction
33	Celebes Hawk-eagle	Is: primeval forest: probably rare	Forest destruction
34	Philippine Hawk-eagle	Is: forest species: little known	Forest destruction
35	Isidore's Eagle	Mountain forests, South America: rare	Forest destruction
36	Spot-winged Falconet	Rather small South American range: unknown	Probably not threatened
37	Fielden's Falconet	Scarce in habitat: little known	Habitat destruction
38	Bornean Falconet	Is: local in Borneo: little known	Habitat destruction
39	Philippine Falconet	Is: local in Philippines: little known	Habitat destruction
40	Madagascar Banded Kestrel	Is: humid savannas: little known	Probably not threatened
41	New Zealand Falcon	Is: forests at high altitudes: scarce?	Probably stable
42	Eleonora's Falcon	Is (many): limited numbers	Persecution (some)
43	Kleinschmidt's Falcon	Small South American range: unknown	Probably not threatened
44	Orange-breasted Falcon	South America: status obscure: apparently rare	Probably not threatened

44 species. 29 island species.
Main threat, forest or habitat destruction on islands. (35 species if any possible threat is perceptible.)
In many species no obvious threat is at present known.

Bibliography

Included here is a bibliography of important references for each chapter. It is not exhaustive on birds of prey, nor does it include every reference consulted in the preparation of this book. Reference to the titles listed, however, will usually lead on to a much wider bibliography for those who are more deeply interested.

CLASSIFICATION AND DISTRIBUTION
Brown, L H and Amadon, D. 1969. *Eagles, Hawks and Falcons of the World.* Country Life. London.
Peters, J L *et al* 1931–68. *Check-list of Birds of the World.* Mus. Comp. Zool: Cambridge, Mass.
Stresemann, E and Stresemann, V. 1961. 'Die Handschwingen mauser des Tagraubrogel.' *J. Fur. Orn.* 101: 4: 373–403
Swann, H K and Wetmore, A. 1924–45. *A Monograph of the Birds of Prey (order Accipitres)* Parts I–XVI. Wheldon and Wesley. London.

HABITATS AND THEIR INHABITANTS
References are difficult to list because of the multiplicity of vegetation types which have been amalgamated by a personal interpretation involving much simplification. General readers wishing to gain a broad idea of ecological zones are referred to the series, *The Continents We Live On* published by Chanticleer Press, New York, including *North America* by Ivan T Sanderson, 1961; *Africa* by L H Brown, 1965; *Europe* by Kai Curry-Lindahl, 1964; *South America* by J Dorst, 1967; *Australia and the Pacific Islands* by Allen Keast, 1966; and Asia, by Pierre Pfeffer, 1968. These are broad and authoritative treatises on the natural history of the areas concerned.

On a rather more elementary level the series published by McGraw Hill *Our Living World of Nature* including *The Life of the Far North* by W A Fuller and I C Holmes 1972; *The Life of Prairies and Plains* by D L Allen 1967, give a good idea of the broad ecology of more specific habitat types. They mention birds of prey characteristic of such habitats in passing.

At a scientific level the following are important references with bibliographies covering smaller regions. (IUCN = International Union for Conservation of Nature).
IUCN. 1974. *Biotic Provinces of the World.* 'Further development of a system for defining and classifying natural regions for purposes of conservation.'
IUCN. 1975. *A Classification of the Biogeographical Provinces of the World.* (M D F Udvardy).
UNESCO. 1973. 'International Classification and Mapping of Vegetation.' *UNESCO Sev. Ecology and Conservation No 6.*
Walter, H (ed) 1976. *Vegetationsmonographien den Einzelnen Grossrann.* pub. G Fischer, Stuttgart.
Papers dealing with specific problems of habitats not referred to later are:
Tundra
Bengtson, S-A. 1972. 'Observations on nesting Gyrfalcons *(Falco rusticolus)* in North-east Iceland.' *Naturfraedingurian* 42: 67–74. Also *Ibis* 113: 468–75. 1971.
Cade, T J. 1960. 'Ecology of the Peregrine and Gyrfalcon Populations in Alaska.' *Univ. Cal. Pub. Zool*; 63; 151–290.
Pitelka, F L *et al.* 1955. 'Ecological Relations of Jaegers and Owls as Lemming Predators near Barrow, Alaska.' *Ecol. Monogr.* 25; 85–117.
Taiga
Neufeldt, I A. 1967. 'Notes on the Nidification of the Red Harrier, *Circus melanoleucus* (Pennant), in Amurland, USSR.' *J. Bomb. Nat. His. Soc.* 64: 284–306.
Temperate deciduous woodland
Moore, N W. 1957. 'The Buzzard in Britain.' *Brit. Birds*: 50: 5: 173–97.
Tansley, A G. 1968. *Britain's Green Mantle.* Allen and Unwin. London.

Temperature moorland belt
McVeau, D N and Lockie, J D. 1969. *Ecology and Land Use in Upland Scotland.* University Press, Edinburgh.
Subtropical mountains and woodlands
Martin, J E and R. 1974. 'Booted Eagle breeding in southwestern Cape.' *Bokmakkierie* 26: 1: 21–3.
Subtropical steppes etc.
Olendorff, R R and Stoddart, J W. 1974. 'The Potential for Management of Raptor Populations in Western Grasslands.' *Raptor Research Rep.* No. 2. 47–85. Raptor Research Foundation. Vermilian. South Dakota.
Deserts
Booth, B McD. 1961. 'Breeding of the Sooty Falcon in the Libyan Desert.' *Ibis* 103. A: 129–130.
Jany, E. 1960. *Proc. XIIth Int. Orn. Congress.* Helsinki 343–52.
Tropical savannas
Brown, L H. 1970. *African Birds of Prey.* Collins. London.
Reichholf, J. 1974. 'Artenreichtum, Haufigkeit und Diversität der Greifvögel in einigem Gebieten von Sudamerika.' *J. Fur. Orn.* 115. 381–96.
Tropical Forest
Alvarez, J B. *WWF Yearbook* 1972–73. Project 610 pp 144–8.
Throlley, J-M. 1975. 'Les Rapaces des parcs nationaux de Cote D'Ivoire. Analyse du puiplement.' *L'oiseau* 45: 241–57.
Tropical montane
Altitude Records:
Laybourne, R C. 1974. 'Collision between a Vulture and an Aircraft at an Altitude of 37 000 feet.' *Wilson Bull.* 86: 461.
Coe, M J. 1967. *The Ecology of the Alpine Zone of Mount Kenya.* Dr W Junk. The Hague.
Hedberg, O. 1964. 'Features of the African Alpine Plant Ecology.' *Acta Phytogeogr. Svecica.* Uppsala 144p.
Aquatic Habitats
'Eleonora's Falcon.' Walter, H. V 19. and Vaughan, R. VI 27.
Towns
Galyushin, V M. 1971. 'A Huge Urban Population of Birds of Prey in Delhi, India.' *Ibis* 113: 522.
Pomeroy, D E 1975. 'Birds as Scavengers of Refuse in Uganda.' *Ibis* 117: 69–81.

ANATOMY, STRUCTURE, AND WAY OF LIFE
Balfour, E. 1970. 'Iris colour in the Hen Harrier.' *Bird Study* 17: 1.47.
Brosset, A. 1973. 'Evolution des Accipiters foresters de l'est du Gabon.' *Alauda* XLI: 3: 185–201.
Brown, L H and Amadon, D. 1969. *Eagles, Hawks and Falcons of the World.* Country Life, London.
Cone, C D. 1962. 'The Soaring Flight of Birds.' *Scientific American* 206: 130–42.
Goslow, G E. 1972. 'Adaptive Mechanisms of the Raptor Pelvic Limb.' *Auk* 89: 47–64.
Haukin, E H. 1914. *Animal Flight.* London.
Kruuk, H. 1967. 'Competition for Food between Vultures in East Africa.' *Ardea* 55: 171–92.
Landsborough Thompson, A. 1964. *A New Dictionary of Birds.* Nelson. London.
Peeters, H J. 1963. 'Einiges uber der Waldfalken *Micrastur semitorquatus*.' *J. Fur. Orn.* 104: 357–64.
Pennycuick, C J. 1971. 'Gliding flight of the White-backed Vulture *Gyps africanus*.' *J. Exp. Biol.* 55; 13–38.
Pennycuick, C J. 1972. 'Animal Flight.' *Inst. Biol. Studies in Biology* 33. Edward Arnold. London.
Pumphrey, R J. 1948. 'The Sense Organs of Birds.' *Ibis* 90: 17–199.
Smeenck, C. 1974. 'Comparative Ecological Studies of Some East African Birds of Prey.' *Ardea* 62: 1–87.
Steyn, P. 1973. *Eagle Days.* Purnell, Johannesburg.
Wattel, J. 1973. 'Geographical Differentiation in the Genus *Accipiter*.' *Pub. Nuttall Orn. Club.* Cambridge, Mass.
Welty, J C. 1964. *The Life of Birds.* Constable, London.
HUNTING AND FEEDING METHODS
Brown, L H. 1970. *African Birds of Prey.* Collins, London.
Brown, L H. 1971. 'The relations of the Crowned Eagle

Stephanoaetus coronatus and some of its prey animals.' *Ibis* 113: 240–3.

Brown, L H. 1976. *British Birds of Prey*. Collins, London.

Craighead, J and Craighead, F. 1956. *Hawks, Owls and Wildlife*. Stackpole Co. Harrisburg, Pennsylvania.

Dunstan, T C. 1974. 'Feeding Activities of Ospreys in Minnesota.' *Wilson Bull*. 86: 74–6.

Goslow, G E. 1971. 'The Attack and Strike of Some North American Raptors.' *Auk* 88: 215–27.

Jehl, J R. 1968. 'Foraging Behaviour of *Geranospiza nigra*, the Blackish Crane Hawk.' *Auk* 85: 493–4.

Monneret, R J. 1973. 'Techniques du chasse du Faucon Pelerin *Falco peregrinus* dans un regin de moyenne montagne.' *Alauda* XLI (4): 403–12.

Muuch, H. 1955. *Der Wespenbussard*. Neue Brchm Bucherci 151 Wittenberg-Lutherstadt.

Rudebeck, G. 1950–1. 'Choice of Prey and Modes of Hunting of Predatory Birds.' Oikos, 2: 65–88 and 3: 200–31.

Snyder, N F R and Snyder, H A. 1968. A Comparative Study of Mollusc Predation by Limpkins, Everglade Kites, and Boat-tailed Grackles. *The Living Bird* 9.

Southern, W E. 1964. 'Additional Remarks on Winter Bald Eagle Populations Including Remarks on Biotelemetric Techniques and Immature Plumages.' *Wilson Bull*. 76: 2: 121–37.

Steyn, P. 1973. *Eagle Days*. Purnell, Johannesburg.

Tinbergen, L. 1946. 'Die Sperwer als roofvij und van zangvogels.' *Ardea* 34: 1–213.

Valverde, J A. 1959. 'Moyeus d'expression et hierarchie sociale chez le Vantour Fauve *Gyps fulvus*.' Alauda 27: 1–15.

Van Laurick-Goodall J, and H. 1966. 'Use of tools by the Egyptian Vulture *Neophron percnopterus*.' *Nature* 5069: 1468–9.

Vaucher, C A. 1971. Notes sur l'ethologie de l'Aigle de Bonelli, *Hieraaetus fasciatus*. *Nos Oiseaux* 31: 101–11

MIGRATION AND NOMADISM

Most of the manned migration stations, such as those at Falsterbo, Heligoland, Fair Isle, Hawk Mountain and Cedar Grove produce periodical or annual reports and/or summaries which should be consulted for details on raptors as well as other birds.

Bannerman, D A and Priestley, J. 1952. 'An Ornithological Journey in Morocco in 1951.' *Ibis* 94: 678–82.

Berger, D D and Mueller, C H. 1970. 'Prey Preferences in the Sharp-shinned Hawk: the Role of Sex, Experience, and Motivation.' Auk.

Berger, D D and Mueller, C H. 1973. 'The Daily Rhythm of Hawk Migration at Cedar Grove, Wisconsin.' *Auk* 90: 591–6.

Broley, C L. 1947. 'Migration and Nesting of Florida Bald Eagles.' *Wilson Bull*. 59; 3–20.

Broun, M. 1948. *Hawks Aloft*. Dodd Mead, New York.

Brown, L H. 1970. *African Birds of Prey*. Collins, London.

Brown, L H and Amadon, D. 1969. *Eagles, Hawks and Falcons of the World*. Country Life, London.

Dorst, J. 1956. *des migrations des Oiseaux*. Payot, Paris.

Evans, P R and Lathbury, G W. 1973. 'Raptor Migration across the Straits of Gibraltar.' *Ibis* 116: 127–34.

Irby, F M. 1934. *Emu* 33: 268–71.

Beaman, M and Galea, C. 1974. 'The Visible Migration of Raptors over the Maltese Islands.' *Ibis* 116: 419–31.

Landsborough Thomson, A. 1942. *Bird Migration*. Witherby, London.

Mead, C J. 1973. 'Movements of British Raptors.' *Bird Study* 20: 259–86.

Mueller, H C and Berger, D D. 1968. 'Sex Ratios and Measurements of Migrant Goshawks.' *Auk* 85: 431–6.

Nisbet, I C T and Smout, T C. 1957. 'Autumn Observations on the Bosporus and Dardanelles.' *Ibis* 99: 483–99.

Osterlof, S. 1951. 'Fiskgfusens flyttning.' *Var Fagelward* 10: 1–15.

Safriel, U. 1968. 'Bird Migration at Elat, Israel.' *Ibis* 110: 283–320.

Snow, D W, 1968. 'Movement and Mortality among British Kestrels 1968 (*Falco tinnunculus*).' *Bird Study* 15: 65–83.

Walter, H. 1968. 'Die Abhangigkeit des Eleonorunfalken (*Falco eleonorae*).' Bonn.

BREEDING BIOLOGY

Balfour, E. 1957. 'Observations on the Breeding Biology of the Hen Harrier in Orkney.' *Bird Notes* 27: 177–83 and 216–23.

Brown, L H. 1952. 'On the Biology of the Large Birds of Prey of the Embu District, Kenya Colony.' *Ibis* 94: 577–620 and 95: 74–114.

Brown, L H. 1955. 'Supplementary Notes on the Biology of the Large Birds of Prey of Embu District, Kenya Colony.' *Ibis* 97: 38–64 and 183–221.

Brown, L H. 1966. 'Observations on some Kenya Eagles.' *Ibis* 108: 531–72.

Brown, L H. 1969. 'Status and breeding success of Golden Eagles in north-west Sutherland in 1967.' *British Birds* 62: 345–63.

Brown, L H. 1974. 'Data Required for Effective Study of Raptor Populations.' *Raptor Research Rep*. 2; 7–20. Raptor Research Foundation. Vermilion, Ohio.

Brown, L H and Amadon, D. 1969. *Eagles, Hawks and Falcons of the World*. Country Life, London.

Brown, L H and Hopcroft, J B D. 1973. 'Population structure and dynamics on the African Fish Eagle.' *Haliaeetus vocifer* (Dandin).

Cadbury, J and Balfour, E. 'Observations on the Biology of the Hen Harrier in Orkney.' *RSPB Research Report*.

Cave, A J. 1968. The Breeding of the Kestrel, '*Falco tinnunculus*,' in the Reclaimed Area, Oostclejk Flevoland, Nederlands. *Irum. Zool*. 18(3): 313–407.

Gargett, V. 1970. 'Black Eagle Survey, Rhodes-Malopos National Park. A population study 1964–68.' *Ostrich* supp. 8: 397–414.

Gargett, V. 1972. 'Observations at a Black Eagle Nest in the Malopos, Rhodesia.' *Ostrich* 43: 77–108.

Grote, H. 1942. 'Einigen urov die Fortpflanzungsbiologie Asiatischen Wurgfalken.' *Betr. Fortpflanzungsbiologie 1942*: 91–6

Gordon, S. 1955. *The Golden Eagle, King of Birds*. Collins, London.

Holstein, V. 1942. Duehögen: Copenhagen.

Horvath, L. 1955. 'Red-footed Falcons on Ohat Woods, near Hortobagy.' *Acta Zool. Aead. Sci Hungariae* 1: 245–87.

Houston, D. 1976. 'Breeding of the White-backed and Rüppell's Griffon Vultures, *Gyps africanus* and *G. rueppellii*.' *Ibis* 118: 14–40.

Livessedge, R. 1962. 'The Breeding Biology of the Little Sparrowhawk, *Accipiter minullus*.' *Ibis* 104: 399–406.

Meyburg, B-U. 1970. 'Zur biologie des Schreiadlers (*Aquila pomarina*).' *Deutsche Falkenorden* 1969: 32–66.

Meyburg, B-U. 1974. 'Sibling Aggression and Mortality Among Nesting Eagles.' *Ibis* 116: 224–8

Ratcliffe, D A. 1970. 'Changes Attributable to Pesticides in Egg Breaking and Eggshell Thickness in Some British Birds.' *J. Appl. Ecol*. 7: 67–115.

Rowan, W. 1921. 'Observations on the Breeding Habits of the Merlin.' *British Birds* 15: 122–9, 194–202, 222–31, 246–53.

Rowe, E G. 1947. 'The Breeding Biology of *Aquila verreauxi*.' *Ibis* 89: 387–410, 576–606.

Sibley, C G. 1960. 'The Electrophoretic Patterns of Avian Egg-White Proteins as Taxonomic Characters.' *Ibis* 102: 215–84.

Schonwetter, M. 1960–6. *Handbuch der Oologie*. W. Meise (ed). Berlin, Akademie Verlag.

Steyn, P. 1973. *Eagle Days*. Purnell, Johannesburg.

Temple, S A. 1972. 'A Portable Time-lapse Camera for Recording Wildlife Activity.' *Journ. Wildl. Management*. 36 (4) 944–7.

Tubbs, C R. 1972. 'Analysis of Nest Record Cards for the Buzzard.' *Bird Study*. 19: 97–104.

Vaughan, R 1972. '*Falco eleonorae*.' *Ibis* 103: 114–28.

Willgohs, J F. 1961. 'The White-tailed Sea Eagle, *Haliaeetus albicilla* in Norway.' Norwegian Universities Press, Bergen.

THE ECOLOGY OF PREDATION

Allison, A C. 1970. *Population Control*. Penguin, Harmondsworth.

Blondel, J. 1967. 'Reflexions sur les rapports entre predateurs et proies chez les rapaces.' *Terre et Vie* 1 : 5–62.

Brown, L H. 1973. *The Mystery of the Flamingos*. 2nd enlarged ed. E A Publishing House, Nairobi.

Brown, L H. and Cade, T J. 1972. 'Age classes and population dynamics of the Bateleur and African Fish Eagle.' *Ostrich* 43 : 1 : 1–16.

Brown, L H. and Watson, A. 1964. 'The Golden Eagle in relation to its food supply.' *Ibis* 106 : 1 : 78–100.

Cheylan, C. 1973. Notes sur la competition entre l'Aigle Royal *Aquila chrysaetos* et l'Aigle du Bonelli *Heiraaetus fasciatus*.' *Alauda* XLI : 3 : 203–212.

Lack, D. 1954. 'The natural regulation of animal numbers.' Oxford.

Lockie, J D. and Stephen, D. 1959. 'Eagles, lambs and land management on Lewis.' *J. Anim. Ecol.* 28 : 1 : 43–50.

Murphy, J R *et al.* 1969. 'Nesting ecology of raptorial birds in central Utah.' Brigham Young University. *Biol. sec.* X : 4, 1–36.

Schaller, G. 1967. *The Deer and the Tiger*. University of Chicago Press, Chicago.

Schaller, G. 1972. *The Serengeti Lion*. University of Chicago Press, Chicago.

Schipper, W J A. 1973. 'A comparison in prey selection in sympatric harriers, *Circus*, in western Europe.' *Gerfaut* 63 : 17–20.

Storer, R W. 1966. 'Sexual dimorphism and food habits in three North American accipiters.' *Auk* 83 : 423–36.

Thiollay, J M. 1967. 'Ecologie d'une population de rapaces diurnes en Lorraine.' *Terre et Vie* 2 : 116–83.

Van Beusekom, C F. 1972. 'Ecological isolation with respect to food between Sparrowhawk and Goshawk.' *Ardea* 60 : 72–96.

Voous, K H. 1969. 'Predation potential in birds of prey from Surinam.' *Ardea* 57 : 117–48.

CONSERVATION AND PROTECTION

Bolen, E G. 1975. 'Eagles and sheep : a viewpoint.' *Journ. Range Management* 28 : 1 : 11–17.

Curey-Lindahl, K. 1972. *Let them Live*. William Morrow, New York.

Fisher, J F, Simon, N, and Vincent, J. 1972. *The Red Book : Wildlife in Danger*. Collins, London.

Fitter, R S R. 1967. *A Dictionary of British Natural History*. Penguin, Harmondsworth.

Koford, C. 1953. 'The California Condor.' *Audubon Soc. Research Report*.

Mellanby, K. 1967. *Pesticides and Pollution*. Collins, London.

Meyburg B-U and Garzon-Heydt, J. 1973. 'Sobre la proteccion del Aguila Imperial *(Aquila heliaca adalberti)* aminorando artificialmente la mortandad juvenil.' *Ardeola* 19 : 107–28.

Ratcliffe, D A. 1963. 'The status of the Peregrine in Great Britain.' *Bird Study* 10 : 2 : 56–90.

de Vries, T. 1973. 'The Galápagos Hawk.' PhD Thesis, University of Amsterdam.

Index

Numbers in italics refer to illustrations.

254